FORCES OF LIFE SHEPARD SLG

FORCES OF LIFE SHEPARD SLG

ANXIETY, DEPRESSION, SELF-HELP, SOCIAL SKILLS, SUCCESS

DAVID LLOYD SHEPARD

SLG

DEDICATED TO ALL OF THE TEACHERS AND PARENTS
WHO TRY TO MAKE THIS A BETTER WORLD.

PSYCHOLOGY; Success and Failure in Ourselves and Our Children. What Can We
Learn From the Greatest Minds in Psychology and the Most Successful People
About How to Increase Success and Reduce Failure?

DISCLAIMER: This book provides knowledge to help us make it through life. Combining education with
therapy can work better than each alone. For serious problems always consult a professional.

CONTENTS

POSTED REVIEWS:

Mendi December 24, 2020

This book is must read if you are struggling with negative feelings or are trying to solve the success puzzle. The author does a great job educating the reader on the primary forces that shape our judgements and how we perceive the world. After laying the foundation of how our brains are wired to think, the author then addresses all of the pinpoints that everyone goes through in life and how to not only overcome them in a healthy way, but how to prevent them altogether. This is an insightful book that also sheds light on the holes in our school system and how to undo the toxic programming we learned in school.

My favorite "aha" moment I had when reading the book is learning how many of us deal with pain in silence to think it's unique to our experience. However, simply learning how common our obstacle or pains are to larger society and seeing proof of how other people shared their experiences and have gone on to succeed really helps calm your mind and you don't feel defeated anymore. Powerful book, the author clearly put a lot of time and effort into perfecting this complex topic and making it easy to process and understand. I strongly recommend everyone read this book, it'll help you improve every aspect of your life!!

Rosetta January 21, 2021

This is a well-researched psychology book. We must change treating problems after they happen into a proper educational science to help prevent the problems. The author puts a great emphasis on children's education. This sentence summarizes it well: The nations who learn "secrets" of how to motivate their children will outclass us in the decades to come.

PAT December 25, 2020

The prologue of the book mentions a statement, "What is it that makes for success and failure in life?" The author as an associate clinical psychologist does a good job of answering this question by providing real life examples and lessons that will enable us profit from others lives to acquire the social skills and knowledge that will make for exceptional success in our personal and professional lives. The last chapter with its 25 points is the best chapter in the book. Recommended reading for all those wish to succeed in life.

Vic December 25, 2020

This book cites multiple entertaining and informative resources to explain why we are who we are. Thought provoking reasoning for the person or family that seeks answers to what make me, him or her click. Most revealing is the aspect of the environment. How we perceive ourselves is the result of our observation on how others perceive us. Special attention is given to how our words matter and can build or tear down. A huge responsibility.

JC December 27, 2020

A well research book about the root causes of success from some of the greatest mind in psychology. The premise of the book stand on the understanding that knowledge and wisdom can be used to created success. He argues that knowledge and understanding of others is the most important key to success in life. The discussion about the different philosophical schools is well expressed and I particularly enjoyed when he contrasted these styles. The book is easy to read and a great reference to learn about success.

Ranieri, January 18, 2021

This book was well-written and very inspiring. It touches on a lot of concepts that have to do with psychology that you can add to your life for more happiness and success. I would recommend this read!

PROLOGUE

ANXIETY, DEPRESSION, SELF-HELP, SOCIAL SKILLS, SUCCESS, PSYCHOLOGY: What is it the greatest minds in psychology, and those who are the most successful in life, know about success and failure? How can we use this knowledge to prevent problems and increase success? Knowledge is the key to success and to surviving failure.

"The only defense against the world is a thorough knowledge of it." John Locke.

"If only I knew then, what I know now..." is the most common regret of life. It screams the critical importance of the knowledge that will allow us to succeed, yet there are few sources to learn from. In their own words, we will read about the experiences that inspired the lives and success of others that help give us a blueprint of life.

Of all the professions, teachers and parents have the greatest chance of changing the lives of others for the better. Yet we all have the ability to improve ourselves.

Knowledge and understanding can change your brain as effectively as any therapy. Preventing psychological problems can be more effective than treating them after they occur. This book is about how learning from others' lives can protect ourselves and our children from the pain of living and learn the social skills and knowledge that make us successful in our personal lives, in raising our children, in our business.

Few people imagine that the most successful and famous people in history have also gone through dramatic problems. From *"Staying Sane in an Insane World"* to *"The Fire in Your Belly"* these are the lessons of life and the skills that make life better for all of us.

How can we develop the social skills we need to succeed in life? How can we avoid the problems life creates for us? Sometimes we forget the problems we have as soon as they are passed, while other times we lay awake at night and agonize over our failures in the past.

EDUCATING people to understand the problems they will face in life, and the skills they need to flourish, is far more effective than psychotherapy after the fact. Yet psychologists do not get paid for preventing problems, only for trying to unravel the knots in your mind after the roadblocks in life overwhelm us.

PREVENTION is always more effective than using therapy after problems arise. Preventing heart disease and stroke by lowering blood pressure is always more effective than treating the problems after they begin. The same is true of psychological problems.

KNOWLEDGE can help to provide answers to what creates success in life, and knowledge can desensitize us to the problems of life that create roadblocks to our success.

SOCIAL SKILLS and UNDERSTANDING OTHERS are basic to success in school, in relations with our spouse, our children, our employees, our business. Yet we do not teach what we need to know in our schools.

J. K. Rowling, the best-selling author of all time, had her manuscript on Harry Potter rejected twelve times by publishers before finding one who gave her a chance. Shy, and feeling defeated, she wrote her first masterpiece while living in a one-room flat, a single mother, divorced, broke, despondent, supported on the government dole.

Jack Canfield, the author of the now-classic *Chicken Soup for the Soul*, says he had been rejected 143 times by agents and publishers before he hit the 144th time. His publisher expected it might sell 20,000 copies. It sold 500 million worldwide.

Most of us give up after failing only a few times.

President Joseph Biden says that as a child he suffered from a severe stutter, so bad he could hardly get a word out without stuttering. What do the other kids do when you stutter? Biden notes they made fun of him, even though they didn't mean to, they humiliated him. That makes it worse.

Finally, he was determined to change. He would practice for hours in front of a mirror, reading long sentences over and over. He learned he had to slow down his speech. Even in his acceptance speech after he was sworn in as President, he showed one reporter how he made notes in his speech where he had to pause, to slow himself down. It is trying to hurry through speaking or, in the case of dyslexia, trying to rush to get it over with is a cause of the problem.

Today, kids who are stutterers write to Biden telling him what an inspiration his story is to themselves.

Challenge yourself. Practice life.
Improve yourself. Re-write your best work, the best writers constantly revise after each failure.
Re-invent yourself after rejection, learn the social skills that few know. Change direction.
Educate yourself. Read. Learn. Practice. Learn what the most successful people know.
Challenge yourself to work to understand other people. Practice new social skills.
If necessary, find new goals.
The scars will take care of themselves

Never think that other people succeed because they are better than you. It is about being able to persist in the face of the inevitable failures of life. And a little luck. And a lot of work. And about learning what you need to know to succeed.

But do not be afraid to change direction. We could name a dozen more people whose arrogance led to failure because they were unable to change direction or go with new ideas or listen to criticism.

After studying why some cultures succeed better than others, famed psychologist Phillip Zimbardo noted that all parents praise their children for *succeeding*, but the ***most*** successful parents praise their children for ***trying***. That gives a child the feeling that it is OK to fail. It may be hard to prove that scientifically, but it is perhaps the most valuable idea for working with children, or adults. But good parents tend to overdo everything. A light touch is best. Keep on, keeping on.

Yet, so many bosses, spouses, and parents have grown up thinking that the only way to change their children or employees or spouse is to be critical of their mistakes, remind them of their failures, get tough to teach them a lesson. That is not what the evidence shows. If children or employees are frequently criticized, they do not get better. They learn to give up; they think, "nothing I do pleases him or her, they only criticize". They lose their desire to do better.

You can get away with being critical with young children, they cannot talk back, but when they become teenagers, the boys may become "oppositional-defiant", and refuse to listen, the girls may become "passive-aggressive", that is, they say they will do what you tell them, then intentionally do just the opposite. Employees who do not feel appreciated may work hard, but only when you are standing over them. That is not the goal you want.

Give them passion. See chapter IV and V on what motivates us all. We can all be inspired by a word from a parent, praise from a teacher, appreciation from a spouse, an idea from a book. Knowledge and understanding can change lives and give us a goal to work toward.

SUCCESS IN LIFE is not about becoming rich and famous, it is about constantly improving. Learn that others are not perfect either, that helps you to be more satisfied with your life. Come to grips with the hurdles others have put in your way. Find peace with the judgmental values others have put in your mind, and learn to recognize the bogus self-judgments we make when we put ourselves down. Get in control of your own mind. After that, work for what you want, never be discouraged by failure or the value judgments of others. Fame is not success.

Famed Civil War general William Tecumseh Sherman, after a furious battle in which his best men and best friend were killed, said, "*We know what fame is, fame is to die in battle and have your name spelled wrong in the newspaper.*"

If you consider the lives of some of the most famous people in history, you will see that fame and money do not bring happiness. Marilyn Monroe, Janis Joplin, Michael Jackson, Britney Spears, Bill Clinton, Donald Trump, and hundreds of others found out that fame and money are not the criteria by which we should judge our lives. The cheers of others can stop suddenly and the fall from such a great height hurts worse when you hit the ground. It is not always the cheering of others that determines our value.

There is a famous video ad put out by Apple computers eulogizing those in our history who have made the greatest contribution to society. It is an ode to Einstein, Gandhi, Martin Luther King, Bob Dillon, Alfred Hitchcock, and so many more, narrated by Steve Jobs himself:

Here's to the crazy ones. The misfits. The rebels. The troublemakers. The round pegs in the square holes. The ones who see things differently. They're not fond of rules, and they have no respect for the status quo. You can quote them, disagree with them, glorify and vilify them. About the only thing you can't do is ignore them because they change things. They push the human race forward. And while some may see them as crazy, we see genius. Because the people who are crazy enough to think they can change the world, are the ones who do.

SURVIVING THE VALUE JUDGMENTS OTHERS PUT IN OUR BRAINS

"HOW YOU MADE ME FEEL" DETERMINES OUR REACTION TO OTHERS

"... people may forget what you say, people may forget what you do, but people will remember how you made them feel." Maya Angelou

Angelou's quote is one of the most important to understand how what we do or say leaves an impression on other people's minds and jades our reaction to others. Our brain is an emotion-driven machine, not a rational one. We all want to feel appreciated by others. We all feel positive when others are positive toward us, and badly if significant others put us down.

If we are unaware of the forces in the environment that affect our minds, then these forces may blow us about like dust in the wind. Understanding how those forces affect our mind is essential to protecting ourselves from the problems of life.

VALUE JUDGMENTS:
What We Need To Know to Survive In the Real World.

Every other thought we have is an emotional judgment of someone or something. The brain is a value judgment machine, constantly evaluating everything in terms of whether it makes us feel good, bad, or angry. As children, the emotional judgments we hear from others become embedded in our minds. We grow up believing in those value judgments. We come to judge ourselves, based on the value judgments we learn from others. Others are often cruel in their judgments of us.

Writer Ernest Hemmingway, who had an image as a tough, hard-drinking, hard-living, man's man, once said that to make it through this life, everyone needs *"a built-in, cast iron, infallible crap detector."* None of us is born with such a device—we have to learn it in the pain of trial and error in life, love, and interaction with others. Or, we could teach it to our kids before problems arise.

In his classic *Walden*, Henry David Thoreau wrote of the great masses of people, *"living out lives of quiet desperation."*

I have been bent and broken, but—I hope—into a better shape." Charles Dickens

"The world breaks everyone," said Hemmingway, *"...and we get stronger at the breaks."* Yet, we only get stronger if learning provides desensitization and counter-conditioning. Learning is the key. Simply being broken is not a recommended outcome.

Psychologists know that girls in our culture start out with a high sense of self-esteem until they hit middle school. Then, their self-esteem goes down. Incredibly, it may not recover for many years.

Why? Because the value judgments of others start in our schools. Let the comparisons begin; how you rate on a scale of one to ten. Do you dress, act, look the right way?

Who is picked first or last in choosing sides in gym class?

Who are the friends you sit with at lunch? Are you sitting with the "cool" kids or the "nerds"?

Are you going steady, or are you left out in the rating-dating game?

It does not even matter if you are singled out for comparison, these ideas are already embedded in our minds, learned by the same subtle process that we learn the language we speak. We learn it from the words and actions of others.

Women are two to three times more likely to be depressed than men, psychologists say. Do not believe it—three and a half times more men commit suicide than women, mostly due to the method used. Men just feel they have to play-act at being tough guys; we can take it, we can't complain, and we can't let others think we are a wuss.

Some kids do very well in our schools, some do very poorly; most just survive with little direction or awareness of what goes on all around us. Those who have good friends may easily deal with problems. Yet even the "popular" kids, the Kardashians of the high schools, have trouble with social skills. If we are alone, the new kid on the block, or never learned the social skills that make life easier, we will have a harder row to hoe to make it through life with success. Without this knowledge, we may continue to muddle through life.

All of us are immersed in a world of value judgments by others. Understanding that *you are not alone* is basic to protecting ourselves from the harmful effects of the judgments of others. Next, we must learn just how bogus the judgments of others can be.

"THE NAIL THAT STICKS UP..."
The Value Judgment Of Our Peers:

Consider the successful motion picture *E.T.* starring an alien with a body like a giant worm, who was marooned on earth. In the movie, children saw him as something of a lost puppy. They befriended E.T. and hid him from the adults in their garden shed. It is a moving story of love and acceptance seen by millions.

Writer Erma Bombeck noted that if this had been real life, the children who first saw E.T. would have run screaming to their parents in fear, and their father would have gone out and hacked it to death with a hoe.

We always prefer our fairytales to our reality. The reality is that we learn early in life to avoid anything that sets us apart from others. Children desperately want to avoid standing out, to keep from being different because they know what happens to those who are. They do not want to become a target of stares, smirks, or laughter. They want to wear the "right" style of hair or clothes, and talk and look the same way—anything to fit in, to be one of the group, to be accepted. They will even pretend they did poorly on exams so they will not be labeled a "nerd" . . . so they will not be made to feel the horrible stigma of being different from the rest.   That, is part of what society loosely calls "socialization".

Those who are different get whacked with the hoe of public opinion.

The Japanese have a saying, *"The nail that sticks up, gets hammered down."*

Any stimulus, even your name, can trigger an emotional reaction in others. Your name does not determine what you are like. Yet, if other people make fun of your name, it is as though they are making fun of you. Your name might not be that big of a problem if you were a two-hundred and sixty-pound, six-foot football player by the name of Dick Butkus. But most of us are not.

A Texas county is named after a former governor of Texas, Jim Hogg. Jim Hogg had a daughter. He named her Ima.

Perhaps he thought he was being cute, but can you imagine the reaction children would give after their teacher went around the room asking them to tell their names?

Urban legends have it that he also had a daughter named "Ura," but that is only an urban myth.

An older man who came back to college, a student of mine, stopped by after class one day when we spoke about this. His name was Oscar. Oscar told me that he went through hell growing up because that was when the Oscar Mayer wiener song was popular in commercials; *"I wish I was an Oscar Mayer wiener... that is what I truly want to be. Cause if I were an Oscar Mayer wiener, every-one would be in love with me."* Really?

Every day in school, it was like,

"Hey, Oscar! Show us your wiener!"

Or: "Hey, Oscar, bite me!"

Or: "Hey, Oscar, Susan wants to ride in your wiener mobile."

The other students laughed.

The laughter of others is what ensures the name-calling will continue.

They thought they were being cute, but it has a price for those who are not aware of what is happening to them. The emotion we remember is" how they made me feel."

Johnny Cash sang a tale of *"A Boy Named Sue."* What happens to a boy named Sue in our culture? Other boys make fun of him and put him down and he gets into fights. In the song, he chases down the father who gave him that name. His father says that he knew he would not be around to raise him, so he thought that with a name like Sue, he would have to get tough or die. Do not bet that he will get tough.

In our schools, even our name could make us like that dog with a broken tail.

PREVENTING PSYCHOLOGICAL PROBLEMS:
Counter conditioning/Desensitization

Our kids are getting even worse attacks from others; it can go on for months. One 14-year-old girl received mean tweets that said, "You are so ugly. Why don't you kill yourself?" She did. Today, you hear of many cases of bullying, and parents have no clue how to prevent the problem. Every year in America some 5,000 teens and young adults commit suicide. Year after year. Yes, really. So, how do you deal with these problems?

One organization that has shown itself to be effective and provides help is the suicide prevention hotline at 800-273-8255.

But our schools and colleges could also help. The key is educating children and adults to understand that they are not alone; other successful people have been through this and gone on to succeed and do well.

You cannot just tell people to "get over it." That never works. But, if you *show* them examples of how even attractive, successful people have been called names, put down, and dumped on, then that can reduce their emotional trauma to cyberbullying.

Jimmy Kimmel shows a series of famous actors reading the hateful tweets they have received in a series you can now find on YouTube. These were stinging bits of hate mail, all directed at beautiful, successful actors and actresses. You may think these are extreme, yet our children hear even worse.

Academy Award and Golden Globe-winning actress Gwyneth Paltrow read a tweet she received: *"Gwyneth Paltrow, you ugly ass big bird looking bitch, shut the f*** up..."*

Writer, director, Emmy Award nominee, and two-time Golden Globe winner Lena Dunham read the mean tweet she received: *"Lena Dunham's boobs are dog noses."*

Model, actress, producer, and movie star Geena Davis, who says she had low self-esteem and felt left out in high school because she was six feet tall, read: *"Geena Davis is a real man's man."*

Beautiful blond actress Chloe Moretz read the mean tweet she received: *"Chloe Moretz, or whatever her stupid name is, looks like my a** hole. Seriously she is not decent looking at all."*

To which she replies, *"You must have a really bleached a** hole."*

Censoring reality is not the answer.

If we show young people that even attractive, successful people can be trashed by others, then that simple experience can counter their loss of self-esteem, and they feel better about themselves. That is good therapy. Even for adults, it helps. But you cannot just tell them, because logic and reasoning fail when up against conditioned emotional stimuli. You have to *show* them. Hence the value of Kimmel's work. Watch it yourself www.youtube.com/watch?v=LABGimhsEys.

Or search 'Celebrities read mean tweets.'

Who would have guessed that Jimmy Kimmel could put out better psychotherapy than psychologists or counselors? Who would have guessed that the same internet that produces problems could be used to treat those problems?

These examples help desensitize us from the put-downs and failures in life. More than this, if we can laugh at ourselves for the things that have bothered us in our past, that marks a sea change in our understanding of life. Yet, when we are young, and have few successes behind us, it is much harder to see any humor in the judgment others dump on us.

Parents can use similar examples on YouTube to protect their children against such attacks on their self-esteem. Of course, we have to be careful that this does not encourage the same behavior.

By the same means, teachers can use the schools that create the problems, to prevent the problems by educating kids to understand the problems others have gone through in life.

It is not enough to tell people "You are great!" Or, "Feel the Power within!". Or, "Be yourself." Or, "Just ignore them". We must *show* people how the words of others have created the problems in the first place, so they will know what to watch out for. Be aware. Learn to untangle the knots others have created in our mind.

Most of the ideas put into our brains by our fellow morons when we are young, or by a society weaned on clichés as adults, have no validity. Yet the force of those ideas on our minds can be overwhelming if we do not have the knowledge it takes to understand.

COUNTER-CONDITIONING, DESENSITIZATION, CONTROL:
Conditioning (learning) is Power

"I do not want to be at the mercy of my emotions. I want to use them, to enjoy them, and to dominate them." Oscar Wilde

Easily one of the most important studies ever done for psychotherapy, and one of the least known, Albert Bandura used the same techniques of Mary Cover Jones that were used to counter condition a child's fear of furry animals.

Bandura went on to discover that he could take *forty* children who were already terrified of dogs, a common childhood fear, and could counter-condition their fear simply by **showing** them 3-minute films of a boy leading a dog on a leash, grooming, petting, and feeding a cocker spaniel. It was a happy boy, playing happily, with a happy puppy.

With only 3 minutes of such films, twice a day for four days, he was able to counter the children's fear of dogs. Bandura is most famous for his modeling studies, like the Bobo doll imitations, but this is his best work. The emotion of simply seeing a film of another child playing with a dog countered the children's fear of dogs, allowing them to lose their fear and pet a dog for the first time.

In only four short days, their fears were overwritten by new learning. Bandura called it, *"A Hair of the Dog."* And it used modeling films combined with desensitization and counter-conditioning.

Simply showing children short videos of a *happy* boy, playing *happily*, with a *happy* dog, could use the emotion of *happy* to counter the emotion of fear.

By the same method, showing them videos of *successful* people who have been trashed by others, and yet are very *successful*, counters the negative emotions of those whose value judgments have infected our minds.

We want children to have a rational understanding of "don't pet a stray dog" or "don't pet a dog that growls." We do not want them to panic. If you run away from a dog, it may trigger an innate response in the dog, and it reacts as if "anything that runs away is prey," and they may attack.

We need to learn to reason, not to fear. Knowledge helps give us that cast iron crap detector Hemmingway spoke of. Understanding the lack of value in the value judgments of others is basic to being able to survive the scars that others leave on our mind.

Improve yourself. Do not let the ignorance and arrogance of others determine your thoughts. Learn from the problems of life the rest of us go through in life that you are not alone. Become an expert in recognizing the bogus thinking of others and our own self-blame.

We all need a life-coach, a mentor, someone to share our success with, that makes success better. Someone to share our pain, that lessens the pain. This book is about how to become your own mind-coach. Educate yourself. Never stop learning.

The whole purpose of life is to make life better for ourselves. You do not have to be a genius, or even a success, at anything except becoming the best you can.

But we can all learn from the success of others what can help us in life. Learning is a key to making life better.

PSYCHOLOGY AS AN EDUCATIONAL SCIENCE
Desensitizing our Fears with Education

Bandura's study would turn out to be perhaps the most important discovery in the history of psychotherapy because it clearly shows that just *seeing* or *learning* stories about other people's experiences can help desensitize our fears, when we see others survive and succeed. It replaces (counters) the emotion of fear with positive emotions. Mary Cover Jones, B. F. Skinner, and Albert Bandura all demonstrated that we can "*Retrain the brain*", although many today ignore giving them credit for the idea.

Anything we can learn; we can change with new learning.

Just as important, Bandura could use this on **40** children at once.

That is something teachers could do in a school system. Not with fear of dogs, as Bandura did, but with the same problems we saw countered by the videos of successful people reading mean tweets they had received. And with the problems of peer group pressure. And much more.

Literature has always been a major source of desensitization to the problems of life, by showing us the problems others have been through, even if fiction. Often literature is the only source available in our schools.

CHANGING THE FORCES OF LIFE:

When I first went to teach at SMU, Dean Joseph Harris told me a story I have never forgotten. After listening to a lecture on how the way others treat us affects our self-concept, some boys in psychology class decided to do an experiment on their own. They selected the shyest girl in class, a girl who sat alone and never spoke to anyone, never said hello to anyone, and, over the next several weeks, they went out of their way to change her self-concept.

It wasn't much. They learned her name and said hello to her by name when they passed in the hall or came to class. They made eye contact. They asked her questions and listened as she spoke, paying attention to what she said. They asked her opinion; What do you think of the teacher? What do you think of the ideas in the textbook? And they showed interest in what she thought.

As the days passed, she started to say hello to them without waiting for them to say it first. She started to ask them questions without waiting for them to ask. She began to think of herself as liked and interesting and popular. More than this, she *became* liked and interesting and popular, and the boys, who started this as something of a lark, now began to compete among themselves to get a date with her. Years later, they would remain friends.

Think of how little it takes to change someone's self-esteem for the better. All it took were words, learning her name, using it to say hello, asking her questions, and listening when she spoke.

The greatest compliment that was ever paid me was when one asked me what I thought, and attended to my answer. Henry David Thoreau

We all know from our own life experiences that we are freer and more capable if we are among a group of people who we feel listen to our opinions and support us. If we are in front of others who judge us, we become inhibited and less able to respond, our adrenaline goes up, our voice may tremble, and our insecurity may show.

Unfortunately, we live in a culture that is often very hostile. Bullying, name-calling, and mean tweets are becoming the norm, and that, too, has a profound effect on others—not just our peers, but teachers and parents are also often judgmental. President Trump used mean tweets to intimidate anyone who crossed him. We will learn how to undo the harmful effects of such bullying and what our schools and parents can do to prevent such harm.

CREATE UNDERSTANDING BY EDUCATION

To give people an understanding of the source of problems there is a Dove film on the corrosive example of the judgments that society puts on girls and women; it is a story of how the effects of rating others have on self-esteem.

Watch this: youtube.com/watch?v=oS8OmSQpb9A

Or search for "A dove film - girl's self-esteem." This is a beautiful story of young teen girls talking about their weight, being told they are ugly (even when they are really beautiful), and not knowing *how to react*, and it fits in, sadly, with what the successful actresses in the "mean tweets" series noted earlier.

She did not know "*how to react*" because we have failed to teach people even what to expect. Make them First Responders, who have been repeatedly trained what to expect, who have a "plan". Give people an understanding. Give them a script.

The ideas others put in our minds are often dishonest, hurtful, and can shape our self-perception for much of our lives. If we simply learn an understanding of how all of us are affected by this dishonesty, and how false it can be, this gives us a beginning of the knowledge it takes to make it through the worst life has to offer. The beginnings of a plan.

In another example, self-esteem and anorexia are beautifully displayed in this video. www.youtube.com/watch?v=uOrzmFUJtrs. Or search for 'Meet yourself: a user's guide to building self-esteem' on YouTube.

In perhaps the most important example, there is an exceptional story of getting over depression in a story by a beautiful and brave young woman named Hunter Kent, who tells her own story. She talks of how she went through serious depression with "self-loathing and hateful" thoughts. She talked about suicide so much, the police were called, as if that were a solution.

She was sent to a therapist, who made her feel worse. That is a story I hear too often from my students. One student called therapy, "Rent-a-Friend." You cannot always depend on counselors or psychologists to know what to do; it is not always taught in our universities.

Hunter was sent to a summer camp for troubled youth. She tells how the people there went out of their way to include her in activities and to make her a part of the group. It did not seem to help.

Then, each student was asked to tell their personal story. Even though she was frightened, she told them of her depression, how bad she felt, her self-loathing, and her history of pain and suffering. To her surprise, all the other students embraced her and became friendlier, and they all became more like a family instead of strangers who were forced to participate. The others could identify with much of what she said. It was the beginning of her feeling better about herself.

Now, she is a beacon of hope to help others, simply by telling her story.

Watch her video; www.youtube.com/watch?v=Rv9SwZWVkOs. Or search for the Ted Talk, 'Conquering depression: how I became my own hero'.

If she had been shown stories such as hers, along with therapy, instead of going through typical counseling, it may have benefited her far more than therapy alone ever could. Better still, we could show such exceptional stories to everyone.

THE MOST EFFECTIVE THERAPY IS...

Her story is an example dramatically similar to the one described by Dr. Joseph Harris. It is the people themselves who have the power to undo the harm and protect the self-esteem of each other. It is the profound effect of other people's value judgments on us, for the better.

The most effective treatment for the problems of life comes from what is often the cause of the problems in the first place, other people; as in the stories of Dr. Joseph Harris or Hunter Kent. People must learn to understand the effect that they can have on others, for better or for worse, and enlisted as a means of preventing the problems.

Sometimes we cannot easily overcome the obstacles to success on our own. Then, we may need help, such as counseling and therapy.

THE SECOND MOST EFFECTIVE THERAPY IS...

The second most effective treatment for such problems is knowledge and desensitization.

Simply learning about the problems others have faced, like the stories on YouTube and Ted Talks, hearing other people tell their stories of pain, and seeing that they have gone on to survive and do well. That removes the terrible feeling that we are all alone, that no one else has ever been there, that there is no hope. Knowledge provides hope.

It is critical to *show* people that they are not alone. Use the examples. Understanding this is the beginning of the end of our self-judgment that creates so much unhappiness. Understanding the words, the mean tweets, that others have imbedded in our brain is essential to counter the speed bumps that society has programmed into our minds.

SHOWING others that all of us go through putdowns, failure, and pain in life is basic to desensitization. It gives us the motivation to keep on keeping on, to jump the hurdles in life, to know we can succeed.

That is what this chapter is all about.

COMBINING EXPERIMENTATION+OBSERVATION

Early on we noted that we need to combine the best of Experimental science with the best of Observational science for the greatest outcome. Combining the experimental science of Mary Cover Jones and Albert Bandura with the Systematic Observation of human problems we have discussed, gives us a tool far better than anything we have in Psychotherapy.

Bandura's work showed he could use video to model learning for forty children to change their fear of dogs. That is the beginning of powerful evidence that using videos, such as those on YouTube and Ted Talks, can desensitize us from our fears of failure, increase our self-esteem, and reduce our tendency to react to peer group pressure.

By showing videos, we can use the existing educational system to educate large groups of children or adults to prevent the problems in the first place.

This is not just for therapy after the fact. If we use the same technique of showing people that others have been through this, *before* people go through it, we can *prevent* the problems before they happen. Knowledge can be power.

Better still, our educational system could use similar examples to educate all children to help each other to prevent this destructive effect *before* it happens. Enlist people to help each other. The other students are key to stopping this, to changing their fellow students' lives for the better and changing our culture.

Yet, nothing is more predictable than that our educational system will do nothing.

Nothing.

Again, it may depend on individual teachers and parents to learn this and use it themselves.

But wouldn't that put psychologists out of business?

Smart psychologists and counselors could use these same techniques in their own practice, even in group therapy. Personal therapy will always be useful. But it can never help more than education can.

TOOLS TO USE...

If you want to see a dramatic example of how anyone can use simple basic behavioral therapy to deal with shyness or lack of self-esteem, go to the personal example of Albert Ellis in chapter VIII on *What We Can Learn From the Great Psychotherapists*. It is an important lesson for psychotherapists, as well as a tool for everyone else. Even business professionals may find this useful, with a little adaptation.

THE NEXT MOST EFFECTIVE THERAPY IS...
Cognitive Behavioral Therapy CBT

CBT may have clients make up a list of what situations they are afraid of, from most to least, and then go up the list desensitizing them along the way. Relaxation techniques, the most effective being what began with Neil Miller's Biofeedback, can be very successful, especially if they actually use Biofeedback.

Or, they have clients make a diary of their good points and their bad points and use reason to point out how irrational their negative beliefs are. Or keep a journal. It works, but it may not be the most effective. It does have an advantage in forcing clients to think about these traits. It can be better than mom telling you, "Count your blessings." Or not.

A common CBT technique is to use reason to convince their clients that your beliefs cannot hurt you if you don't let them. This may be an improvement over our grandmother's saying "sticks and stones may break your bones, but words can never hurt you." Or not.

So much depends on the skill and personality of the therapist, perhaps more than technique.

Yet reason fails badly when faced with an emotionally held belief. We saw this in dramatic form in the Presidential election of 2020 when those whose candidate lost, refused to believe he had lost. Out self-judgment may be so emotionally implanted in our brain, that reality cannot easily change it. All of us tend to get caught up in emotions that can trump reason.

If therapists used the methods of showing examples, as discussed earlier, it could make their therapy far more effective.

THE LEAST EFFECTIVE THERAPY IS...

By far, the least effective therapy is to try to use reason to convince them that the pain they feel is not so bad (in my day...things were so much worse", or "You have it so easy in life today.").

Or, just as useless, telling them to "count your blessings" ("Look at all the things you have to be thankful for.") does not help when up against the value judgment of their peers. Yet this is all most of us know to try to help others. Reason can never compete with emotion in undoing the harmful effects of the value judgment of their peers, or worse, their own self-judgment which comes from the ideas they learn from others.

USING OUR SCHOOL SYSTEM TO PREVENT PROBLEMS

Even the most tested psychotherapy, Cognitive Behavioral Therapy, has often forgotten the origin of the work of Mary Cover Jones (hands-on with little Peter) and Albert Bandura (on video, with 40 kids, with desensitization). That is what works best.

Again, Bandura's work suggests this method could easily be adapted to use in the school system itself, with large numbers of students at a time, or in smaller group therapy meetings.

Education, *by showing them real-life examples*, is far more effective than talk therapy. Teachers and parents could be more effective than psychotherapists, if we gave them the tools and ideas, they need to be successful.

Even more effective than therapy is... prevention.

PREVENTIVE PSYCHOLOGY, IS MORE EFFECTIVE THAN THERAPY

If we show them such examples *before* they are exposed to mean tweets, then it helps to inoculate them against such loss of self-esteem. That kind of preventive psychology is vastly more effective than waiting to do therapy after society has labeled them with negative emotions. Preventive psychology can be far more effective than psychotherapy. But, if our schools do not take advantage of this, and use it to educate *all* our kids, then we continue to fail them.

Everything we have discussed applies to adults, as well; the underlying principles are the same—angry bosses or parents who grew up believing they are justified in their anger to "teach them a lesson." Others who only see the view inside their own heads; bosses who never went to "boss school" or learned how to be an effective, understanding supervisor; unhappy parents who never learned any child-rearing skills; diverse coworkers who view us without understanding; our spouse, children, or all life's forces that make life more difficult for all of us.

The more we understand others, the better off we are in life.

OUR PAST IS PROLOGUE TO OUR FUTURE:
Our Past Experiences Causes Us to *Anticipate* the Future

There was a girl in high school named Bobbi Boydee (not her real name, but a similar example). Bobby with an "i" supposedly makes "Bobby" a girl's name. Her parents thought it was cute because it tripped off the tongue. All through school, others teased her with, "Hey Bobbi Boy, are you a boy or a girl?" Or, "Here comes Bobbi Boy!" She hated her name, and I think she hated herself, too, because of the way others reacted to it.

She became defensive, reacting to others in *anticipation* that they might put her down, before they even said anything. When she got old enough, and had the money, the first thing she did was to go before a judge and have her whole name legally changed.

We often have no clue how words can become embedded in other people's minds. The effects become a part of our self we may try to rid ourselves of.

A similar example may be what happened to singer Michael Jackson. Michael had repeated surgeries on his nose, to the point where it almost could not be restored. Why? Is there a gene for wanting a small nose? Unlikely.

We know Michael's father repeatedly called him "Big Nose," even though he had a perfectly normal nose for a young black man. We know he was sensitive to this, apparently to the degree that he had repeated surgeries to reduce his nose size when he could afford it.

When Michael died, he had pointedly left his father out of his will.

Outstanding singer Karen Carpenter had a problem with anorexia and bulimia. She saw herself as "fat," even when she only had a tiny amount of fat. To rid herself of her fear of fat, she starved herself until her electrolytes were depleted, and her heart gave out.

Psychologists have a name for this—Body Dysmorphic Syndrome. Sticking a label on people does not help, and it does not tell us anything about the cause or cure.

This is not just something that happens to those whose names are different. Anything that makes you different makes you a target for prejudice. A child with different skin color, race, or religion will be signaled out for name-calling if he is a small minority. Today, we might take action if that happened; fifty years ago, we would have ignored it. Prejudice is not just about skin color.

The same prejudice may happen to a shy child, one who acts differently, one who is too talkative, has no friends, is handicapped, or who simply makes a mistake or is different in any way. They will be subjected to stares, questioning, and name-calling.

THE EMOTIONS OF OTHERS LIVE WITHIN US

Emotions live within us, embedded thereby experience, without our awareness. The emotions are established in our brain just like the language we speak; in a manner so subtle, so without our awareness, that we simply do not notice what is happening. Understanding the origin of these emotions that flit through our brain becomes a first step in gaining control over our minds.

Name-calling is often dismissed as "kids will be kids" by school officials. No. Sticks and stones may break your bones, but words can kill you. This is our culture's version of a voodoo curse—in psychological terms, a conditioned emotion; they came to see themselves as worthless from the emotions others use to label them.

There is no difference between racial and religious prejudice and what goes on in school—the put-downs, bullying, and laughter. School is the training ground for a lifetime of problems. The prejudice that leads to bullying and name-calling in children in schools is absolutely the same as the prejudice that leads to racism and religious bigotry.

Studies by Public Broadcasting's NPR show that the vast majority of school shootings, 17 of the original 19 shootings, were committed by kids who were targets of school bullying. They did not just take up a gun to go after the bullies, but often went after everyone, even the ones who laughed, or only stood and watched.

For the kids who take up a gun against the bullies, it is like another cliché in our culture, the "Get even." mentality in our movies. The John Wick Syndrome. "They killed his puppy, now it's "Get even" time."

Yet we label the kids who shoot up a school as, "mentally ill" and blame it on mental illness. And we do nothing to change the causes. Society never blames itself for our failure to protect our kids.

One school principal noted that, following the school shootings; *"For the rest of the year, everyone was really nice to each other."*

The "fear of what other people will think" is, in large part, what society means when people say kids need to learn "socialization" in our schools. Some of that may be useful, we need to understand what other people think. We need to learn to behave with a greater understanding of how our behavior affects others. But, like our reactions to a stray dog, we need a rational understanding, not a fear reaction.

One way or another, anger and hate have made their way into every aspect of our schools, politics, tweets, and news. Yet, the media has largely ignored the cause, even when someone takes a gun and shoots up a school. Instead, we talk about more police security in our school, and harsher sentences for school shooters. How harsh a sentence can you give to a school shooter who kills himself at the end?

Our society has failed our children. And our adults as well.

Problems are not rare events in human experience; problems are the rule. In the past, such problems have been censored by our schools, press, and parents, giving the impression that the problems of living were rare.

These are not "mental illnesses" —they are the norm for human experience. They differ only in degree. The greater the degree, the more it interferes with our life. The greater the interference with our life, the more likely it is to get a diagnosis.

To a greater or lesser degree, we all go through problems in life—they are part of human existence. Sometimes, we forget the problems we have as soon as they are over, we do not have time to dwell on them. While other times, we cannot sleep at night as we agonize over our failures in the past, and try to understand where we went wrong.

THE SKILLS OF SOCIAL INTERACTION: WORDS, THE POWER OF SOCIAL STIMULI

Imagine you pass someone you know in the hall at work or school, someone you have been friendly with before, and you say "Hi! How are you?" and they ignore you. What happens if they just walk by, saying nothing? You put your effort in and got nothing. How do you feel? Maybe you want to shake them, or perhaps you think they are angry with you. If this happens more than once, your behavior will go into extinction—you would never again say hello to them.

We never notice how powerful this simple stimulus-response social expectation is until it is no longer there.

Our mothers taught us to say "please" if we want something and "thank you" if we get it. How would you feel if you went out of your way to do something for someone and they did not bother even to say "thank you"? Most likely, you wanting to help this person would go into extinction.

These are essential skills in social interaction. We learn them so early and so effectively that we do not even think about them; it becomes an automatic behavior.

Some of us are too afraid of what other people will think of us to do well in life. If you are shy, others often judge you as unfriendly, even though that is rarely true, and they avoid talking to you.

For others, verbal combat between spouses, teens, adults, friends, and nations too often ends badly because we never learned to understand how our words affect other people's minds.

As Academy Award-winning actress Kathy Bates said when describing her new role in a Walter Scott interview: *"I try to be diplomatic, but sometimes pterodactyls fly out of my mouth."*

The Great Lady of cultural anthropology is Margaret Mead. Mead is one of the most successful individuals in the history of anthropology, with many books to her credit. Yet, she was never so successful when it came to social skills.

When Margaret was talking on the phone and came to the end of the conversation, she would just hang up. No "nice talking to you," or "I'll see you later." She just hung up.

When she spoke to her associates at Columbia and found she had nothing left to say, she would turn and walk off. Not even a, "We must have lunch sometime" (Yeah, right).

People thought, "That is one rude bitch."

We judge others without understanding why they are the way they are. Margaret never learned some of the basic social skills we all take for granted. Her parents seemed to believe that, like the Summerhill School that says it is a mistake to force cultural values on children, that they should be allowed to grow up "naturally" without our culture's bias. Margaret was never taught these simple skills.

"I was brought up to believe that the only thing worth doing was to add to the sum of accurate information in the world." Margaret Mead

The early ideas from her parents may have given her an advantage in her study of anthropology, but it was an albatross around her neck when it came to dealing with other people. Margaret herself said of social decorum, *"I have a respect for manners as such; they are a way of dealing with people you don't like."* Three times married, three times divorced, Mead was an outstanding scholar but often had problems dealing with the most basic social skills.

We rarely meet someone who grew up without the most basic skills. Few of us would be as curt as Mead, but all of us, even the Kardashians, have problems with social interactions, especially when we do not understand how these stimuli affect others.

We are never forced to seriously consider the truth of our value judgments we make about others. We rarely get to know another person well enough to find out that we are wrong. So, we continue to judge others, and ourselves, with little evidence that our judgment is true or false. We never question the value judgments that others put in our minds.

LEARNING SOCIAL SKILLS

The stimuli we emit may determine our success or failure in life. Some, such as President Truman, were widely known for being curt with others and having few social skills.

Democratic presidential hopefuls John Kerry and Al Gore were often viewed as too intellectual and stuck up, whether true or not. They both lost to George Bush, in part because he came across as a likable, average guy. How did that work out?

Hillary Clinton was seen by many as cold and aloof because of her persona, and probably the picture the media kept running of her with her lips pursed, like an angry school teacher. Everyone told her to smile more; now, she does. Now, people criticize her for "faking" her smile.

Donald Trump is, well.... How did that work out?

Failure is part of life. Being disliked is part of life. The solution is to *understand* this and use your mind and passion to push beyond the inevitable failures that all of us have. But always improve yourself.

Whether we like or dislike another depends on our past history of experience, or lack of it, more than anything else. Most of us are unaware of the simple cues that trigger other people's reactions to us. And the cues we use to judge other people are often very superficial. Yet because we react so strongly, these simple cues are very powerful.

Benjamin Franklin was an exception who was liked by almost everyone. When Franklin died, France declared a day of national morning. You see the same likeability factor in Bill Clinton; not that everyone agrees—even the best-liked people may be hated by some. Even old Ben Franklin was often criticized for being a "womanizer" because he liked to flirt with the ladies.

Those who are the most successful in life, business, school, family, and interpersonal relationships are often the ones who have learned these skills. People skills are basic to every aspect of life. If we fail to learn those skills, we are like that dog with a broken tail.

None of this is what will make you popular. Recent studies suggest that today, those kids in school who trash others are often the popular ones; putting other people down gets a laugh. Other kids think they are "cool." It might get you elected President.

Our educational system has failed us by letting children condition each other. The kids decide what is "cool," and they pass it on to the younger kids with words, words associated with emotion. But it is important to educate them to understand other people well enough so that they can gain control of their own mind and understand why others react as they do, and maybe make life a little better for others and ourselves in the process.

WHAT WORKS IN RELATING TO OTHERS?

Benjamin Franklin's biographers note that when he was ambassador to France, the women of Paris fell in love with him. Why? If you look at his picture on a 100-dollar bill, you see a man balding on top with long hair down to his shoulders and a rather large paunch.

How could he have been so loved? In part, just because he listened to them; he hung on every word because he was sincerely interested in people. In simpler terms, he made eye contact, and he asked questions about what they thought. He did not criticize or make fun of their ideas. He told them stories and often made fun of himself. These are skills many never learn. Whether you are a kid in school or a business professional supervising other people, these are the most basic lessons of life for understanding others and being a success.

Asking questions, making periodic eye contact, and really being interested in others, are valuable social skills by themselves. But those are not the only important social skills Franklin had; he also had a unique social skill that almost no one else has.

The story is told of how, as a young newspaper publisher, Franklin used to be very caustic and critical of people in the press. He made fun of other people's ideas, and he ridiculed their behavior, especially political figures. The people he ridiculed became outraged at him. He offended many.

Franklin described the insight he gained from his experiences in a crisp statement: "*I resolve to speak ill of no man whatever, not even in a matter of truth; but rather by some means excuse the faults I hear charged upon others, and upon proper occasion speak all the good I know of everybody.*"

"Speak ill of no man... excuse the faults of others... say all the good I know of everybody." Now that is a powerful insight. Almost everyone can see that it is a profound social insight, yet almost no one ever practices it.

Just before the second Gulf War, a study was carried out to see what ordinary people were discussing at the office or over lunch. They found that the number one topic of conversation was not the coming war or even the poor economy. Seventy percent of conversations were about other people; people they knew at work, such as their bosses and coworkers, and their friends or spouses.

And seventy percent of what they said about other people behind their back was something negative.

Most of us would have nothing left to talk about if we did not talk about others. We rarely say anything nice about others behind their back. We say things behind their back that we would never say to their face. But if we hear that they said something nice about us to others when we were not there, we take that as a far more meaningful compliment.

"*The nicest feeling in the world is to do a good deed anonymously and have somebody find out*" Oscar Wilde

There is a simple dictum by Robert Louis Stevenson, who said, *"There is so much good in the worst of us, and so much bad in the best of us, that it ill behooves any of us to talk about the rest of us."*

That is a remarkable comment, even a profound idea, but that is not the reality embedded in our minds. Understanding Stevenson's logic is not so easy. Those who understand this have an advantage in understanding human nature and the people around us compared to those who do not.

"It's a waste of energy to be angry with a man who behaves badly, just as it is to be angry with a car that won't go." Bertrand Russell

Even the worst of people do not blame themselves for their failures. They tend to blame others, and pointing out their failures is not likely to make them better—it may just create resentment. If couples learned this lesson early, there might be less divorce, unhappiness, or problems with their children.

That doesn't mean we should let anyone take advantage of us—sometimes we must stand up for our own rights and, better still, learn to argue more effectively, albeit with less anger. People who put us down we should avoid for the sake of our self-esteem.

"Always forgive your enemies; nothing annoys them so much." Oscar Wilde

Today, young people grow up knowing, perhaps from late-night comedians or politicians in the news media, that if we cut someone else down, if we ridicule them in public, then we get a laugh from other people. The comedians get a laugh when they trash a celebrity or politician. Comedians made fun of Britney Spears shaving her head; "Britney's having a breakdown."

Politicians get immediate attention from the press when they trash their opponent. Donald Trump sucked all the oxygen out of the room when he trashed his opponents; the press covered almost nothing else. Kids see this, and they try it in school. They put another kid down or call someone names. The other kids laugh. It works. It is that reaction from others that ensures this behavior will continue. It is the laughter of his supporters that ensured Donald Trump could not stop putting others down, blaming others for his faults, even in the face of a deadly virus.

We grow up in a quite different reality from that of Franklin. It is built into our society, fire formed by our media, politicians, and schoolchildren. It is not natural; it is learned. It is the unintended consequence of our failure to consider what putting our kids in jail for five days a week, eight hours a day, for 12 years would do to them.

ADVICE FROM THE PAST

The journalist Rudyard Kipling wrote an epic poem on life in the real world that holds as true as any advice:

> If you can keep your head when all about you
>> Are losing theirs and blaming it on you,
> If you can trust yourself when all men doubt you,
>> But make allowance for their doubting too;

If you can wait and not be tired by waiting,
　　Or being lied about, don't deal in lies,
Or being hated, don't give way to hating,
　　And yet don't look too good, nor talk too wise:

If you can dream—and not make dreams your master;
　　If you can think—and not make thoughts your aim;
If you can meet with Triumph and Disaster
　　And treat those two impostors just the same;
If you can bear to hear the truth you've spoken
　　Twisted by knaves to make a trap for fools,
Or watch the things you gave your life to, broken,
　　And stoop and build 'em up with worn-out tools:

If you can make one heap of all your winnings
　　And risk it on one turn of pitch-and-toss,
And lose, and start again at your beginnings
　　And never breathe a word about your loss;
If you can force your heart and nerve and sinew
　　To serve your turn long after they are gone,
And so hold on when there is nothing in you
　　Except the Will which says to them: 'Hold on!'

If you can talk with crowds and keep your virtue,
　　Or walk with Kings—nor lose the common touch,
If neither foes nor loving friends can hurt you,
　　If all men count with you, but none too much;
If you can fill the unforgiving minute
　　With sixty seconds' worth of distance run,
Yours is the Earth and everything that's in it,
　　And—which is more—you'll be a Man, my son!

Revise. Reinvent. Learn. Practice. Knowledge provides tools. Persistence improves the odds.

WHAT STIMULI DETERMINES WHO WE LIKE?

A perhaps surprising new discovery in the psychology of what stimuli make other people like us is the discovery that, while talking to someone, if we look them in the eye and we just lightly and briefly touch them on the arm, that light touch dramatically increases the connection we make with another. It makes the connection more personal and more real, and it is a rare skill.

Joe Biden was a candidate who was so well-liked that, in the 2020 election, the Republicans had to attack him by attacking his son Hunter Biden. Guilt by association, even though there was never any evidence against his son.

Joe Biden naturally has practiced the politics of touch for decades. It has served him well. But it also got him into trouble when others felt he was too touchy or familiar. Not everyone sees this the same way, although women can more readily get away with this than men; nothing works the same for everyone.

We have already learned something about how the stimuli we emit will affect the perception of others, and how others will see us. We all try to emit those stimuli we have come to see as positive in other people's minds and avoid what stimuli we see as negative.

Consider all you do just before you go out on a date *before you even leave the house.*

Let's pick on the guys first. Men, before you even leave the house on a first date, what did you do to control the stimuli you emit?

Did you take a shower? Of course. Did you use deodorant? Yes, because you do not want her to know what your pheromones really smell like.

Did you shave? Probably, so she won't think you are the slob you are until after you get married.

Did you wash your car? For sure, because you wanted to impress her with how cool your car is.

Do you pretend you like chick flicks, even though your mind is on *John Wick Chapter 6*?

And when you walked her to your car, do you open the door for her? Do they still do that, ladies?

Then, where do you take her to eat—to the 99-cent special at

Hamburger Patti's Burger Barf? No, because you are saving that for after you get married and have three kids and can't afford anything but the 99-cent special. So, you take her to someplace special that costs more than you want to spend, to impress her.

And ladies, what do you do before you ever leave the house for the date? Do you take a bath? Of course.

Do you shave? Yes, because you do not want to be remembered as Porcupine Patty, the girl with the hairy legs.

Do you pretend you like football and *John Wick Chapter 6*, even when you prefer *Fried Green Tomatoes* and love stories?

And what do you wear? Do you try on one dress—no, that's not right—and a second dress—no, not quite, it has to be just the right one, something in between Hillary Clinton and Lady Gaga. Because you want something that says, "Look at this, big boy," but not one that says, "Take me, I'm easy."

And when he takes you to that fancy restaurant, what do you order? "Oh, I'll just have a salad." Like, "Oh, I hardly ever eat *food*." Or, "I just love the tangy taste of E. coli in the evening."

None of this makes for success, but clearly, we are all trying to emit stimuli that up the odds. We are all dimly aware that the stimuli we emit determine how others see us, but we rarely learn more than we learned from our mothers.

None of these stimuli are as important as the ones noted by Benjamin Franklin. The simplest possible stimuli, ones that do not cost us more than a few words, are the most powerful. Yet few practice the most basic methods. We have been broken by our experiences with our peers in our schools, by what we see on television, by our interactions with others. It is hard to change.

Much of the positive things we do in courtship, will change after marriage; after we sign on the dotted line. Does he still open doors for you, ladies, if he ever did? Does she still pay rapt attention to whatever he says, as if she actually cared? Not likely—which may be one of the reasons why marriages go into extinction.

Women are usually always better at social skills than men are. They learn how to flirt without seeming to; they make eye contact, ask questions, pay rapt attention, and flash a smile. Their early education came from following their mothers around, they subtly picked up these skills by the same way we all learn the language we speak.

It works for men, too, but men hardly ever learn the power of these simple stimuli; we are handicapped by the fact that much of our early education came from the other know-nothings among our male friends. Instead of learning social skills, we play baseball with "the boys." Instead of learning to understand others, we pick up our peers' trash-talk. We start out in life with a broken tail.

The failure of men to have learned such basic skills may account for many cases of divorce, losing a job, not being promoted, or failing in business.

Any teacher will tell you that female students are generally the best students. They pay attention in class, they listen, they make eye contact with the teacher, even if their mind in Cancun.

Women tend to work harder. It is not always true, of course, not by any means, but it is so obvious to teachers, it is hard to miss. Males just want to be entertained. It is a gross generalization, but females are more serious. Males who do not learn such simple social skills, are at a disadvantage in life.

Women also have to learn how to discourage suitors without seeming to; never smile, look away, say you can't go out because you have to stay home and do your hair.

Sometimes, a touch on the arm is too effective at establishing a relationship.

One student, who was trying to discourage a guy pestering her for a date, said, "Why can't men take a hint?!" But men grow up in a different reality. Guys rarely use hints—we are usually more direct; hinting is not an idea guys learn in the male subculture.

EDUCATION: IF ONLY I KNEW THEN...

Throughout our lives, there is one regret that all of us will constantly repeat, "If only I knew then what I know now!" That simple statement screams the importance of knowledge.

If only I knew... I could have been a better friend, a better spouse, or a better parent; I could have been nicer to my parents or spent more time with my kids; I could have been a better person, I could have been a success...

We all recognize, looking back on our lives, how knowing what we know now could have made life better. That is why we need to teach what adults have learned from the pain of living to our young. That is why learning is so important, even to adults. We can never learn enough about how to understand other people.

Yet, we do little to teach our children to learn from our mistakes. We do not give them the knowledge it takes to succeed. We do not teach them how to be better parents than we were. Our schools teach reading, writing, and arithmetic—knowledge that may make them useful as employees working for "the man" in a corporation, but we teach nothing about what makes for success and failure in their own lives.

J. R. Duplantier once described our educational system with its "back to basics" education as designed to produce "Good little worker bees" who will work hard for the man, to fit into corporations, to make rich people richer. "*Good Little Worker Bees*" is now the title of a rock group. Yet, our schools teach little about what they need to know to survive in the social reality of life.

"*I have never let my schooling interfere with my education.*" Mark Twain

We need to teach life lessons, the social skills of living, the childrearing skills our parents never learned, and the knowledge it takes to understand the real world. We need to learn to understand our mind, our emotions, the origin of our ideas, and the relativity of our value judgments and judgment of others.

Moreover, we must learn the skills we need to control our minds. If we fail at this, we will handicap the next generation as well as ourselves. We will continue to produce generations with broken tails.

KNOWLEDGE ALLOWS US TO PREDICT THE FUTURE

One hundred and fifty years of schooling has changed our culture beyond recognition. In the year 1900, the average American never went beyond eighth grade in school. Today, by law, our students are locked in school for eight hours a day, five days a week, for twelve years. No one ever thought, "How will that change our children?"

Society putting its children in schools to be educated is like a dog chasing a car; we never stop to consider the problems that may result if we get our teeth caught on the bumper. By puberty, our kids' allegiances have already changed from their parents to their peers—the peer group rules.

We could have predicted this easily. We could have realized that the kids themselves would become the major force in determining their thoughts, ideals, and views of sex, culture, and what is important in life.

And what do most teens in our schools' value? Not education, not knowledge, not understanding. They value sports, gossiping, and socializing. The whole intent of education is seen by most as something they have to muddle through.

The school system itself has changed society irrevocably. It is the emotional conditioning of the peer group, by the peer group, for the peer group. Then came TV. Then came the iPhone. Then came free back-to-back porn. Then came mean tweets. Then came Facebook and Russian bots in our politics...

"*People will come to love their oppression, to adore the technologies that undo their capacity to think.*" Aldous Huxley

If only we had the good sense as a society, as adults, to get together and consider what we needed to do to prevent the problems of life our children would face, we could have done something to prevent their problems.   We could have realized that bullying, name-calling, peer group pressure, sex, interpersonal relationships, and more would become major problems in their lives. We would have seen how our school system is the prison that forces value judgments on our kids through a peer group with no idea of what they do. We could have taught them what to expect in life.

We did not.

They learned only reading, writing, and arithmetic—even that we did not teach well, in large part because the students themselves saw education as only one more thing they had to endure, while the thrill of sports and gossiping with their friends became what they saw as really important in their lives.

We cannot depend on the system we have now to help us. We are on our own. Perhaps individual teachers and parents can make a difference.

BEYOND WHAT SOCIETY PUTS IN OUR HEADS

"The things other people have put into my head... do not fit together nicely, are often useless and ugly, are out of proportion with one another, are out of proportion with life as it really is outside my head." Kurt Vonnegut

Later in this book, we will take a trip through the minds of the most successful and creative people in history, from Angelou to Einstein, from Oprah to Vonnegut, to see what bits of wisdom we can find to help us understand what is important in life. Along the way, we will discover how our greatest heroes, like Abe Lincoln, went through severe periods of depression and anxiety.

WHY DO PEOPLE FEEL SO BAD THAT THEY WOULD CONSIDER SUICIDE AS A WAY OUT?

The myth that people who commit suicide are "weak" persists in America. Another myth says they are only trying to get attention. It is a national illusion. The major reason why people commit suicide is that they have had, on average, **four times more negative experiences** over the last years than the average person. They see suicide as a way to end the pain of living. You see the same pattern in anxiety disorders and major depression and PTSD; all are preceded by negative experiences that set them up for the "snowball from hell" effect. The show ball gets bigger as it rolls downhill and picks up more snow. Psychological pain can be more unforgiving than physical pain and last longer because our brain repeatedly chews over our failures in the past.

"...there will be better and happier news one day, if we work at it." News Legend News Edward R. Murrow

Apparently, we are not working at it.

One organization that has shown itself to be effective and provides help is the suicide hotline at 800-273-8255.

LEARNING SOCIAL SKILLS

One example of how to change your life for the better is Dale Carnegie, the famous author of an all-time bestseller, *How to Win Friends and Influence People.* Carnegie described himself as a young man with an inferiority complex—he was painfully shy. His mother encouraged him to take classes in public speaking in school to help him get over his problem.

Over the years, Carnegie interviewed many people who had succeeded in life and learned some very important lessons from them about success in interpersonal relationships. He found that successful people tend to attribute their success to their understanding of others and how to work with them.

After many years of teaching others how to speak in public, he finally went public with his magnum opus in 1936. This is an old and dated book by today's standards, yet parents and students often ask me how to teach social skills to their children or learn themselves. One of the few books I have seen that does this is Carnegie's. You can find it in any library, yet we still do not teach young people what Carnegie knew, and few understand what he spoke of so many years ago.

Carnegie spoke of simple things. Learn someone's name, and use it when you talk to them. Understand that criticizing others only makes them defensive—it does not help them. These were the great lessons learned by Franklin that few today understand.

Nothing in life comes easy. However, humans can use their minds to learn the skills, understanding, and goals that are worthwhile in life. We can become aware of what is important to make it through life and what is tripe, put in our minds by society.

INTO THE FUTURE OF PREVENTIVE PSYCHOLOGY: TOWARD PSYCHOLOGY AS AN EDUCATIONAL SCIENCE

The future of psychology has to be in preventing the problems in the first place, not just in treating them after they have formed. Moreover, this has to be a science that benefits everyone, not just people with diagnosable problems. We must change from the idea of being a science of treating problems after they happen and into a true educational science that can help prevent the problems; to fail to use education for prevention is an enormous failure of psychology.

"The best way to predict the future, is to choose it." Abraham Lincoln

We can use the existing educational system to protect our children from the forces that create the problems and teach an understanding of others that can increase their chance for success in life.

In the same way that we can prevent medical problems, such as heart disease and stroke through diet, or prevent cancer by protecting young people from psychological peer group pressure that leads to smoking, we can also prevent psychological problems by simply giving people the knowledge and skills needed to make it through life.

It is far easier to prevent problems from happening in the first place than to treat these problems after they occur. Yet, all of psychology and psychiatry are devoted to treating problems after they occur, unscrewing you after society has screwed you up, and not preventing the problems before they arise.

PREPARE LIKE A FIRST RESPONDER. Train. Practice. Make a plan. Learn from the lessons of others that bad things happen to all of us. Use this understanding to prepare for the problems that will come. Steel yourself. Plan how you will react. Learn to use what you now know to remind yourself, over and over, you can overcome. You are not alone.

This is a book about understanding how forces in the environment control our thoughts, lives, and society, and what we can do to change it for the better. Knowledge and understanding are as powerful as any form of therapy—more important than categorizing thousands of disjointed studies in a textbook and vastly more important than coming up with lists of symptoms to label problems in the DSM-V. We need to learn in school what we need to know to make things better for all of us.

College students flock to our psychology courses, eager to learn the secrets of life. Psychology is easily the most popular of all electives for students; it blows away any other elective—no other subject comes close.

Yet, psychology textbooks provide no clue. Filled with thousands of seemingly unrelated studies, authors studiously avoid giving any help, lest they fall prey to the dangers of the "law of unintended consequences." So, students leave our courses without the knowledge they need to survive the realities of life.

Instead of providing insight into our problems, textbooks only provide thousands of experiments, steeped in obscure language, leading to no clue about real life. If we took a page from the works of Jane Goodall, who used another method of science—Systematic Observation—to show how chimpanzees live in the real world, or the work of Margaret Mead, who used the same techniques to illustrate the different lives of other cultures, we might have been more relevant.

This is a scientific method called Systematic Observation or Naturalistic Observation, and it is a more important way of dealing with the problems of life than laboratory experimentation. But if you can combine both, like combining Albert Bandura's study with the knowledge we have been discussing, it would make it even better. If we could use all of this in our education system, we could make something truly useful out of this.

Surrounded by an immense wealth of human behavior, psychology chose instead to go with thousands of obscure scientific experiments because it wanted to look more "scientific." It has been a great failure.

"We have lost the art of living, and in the most important science of all, the science of daily life, the science of behavior, we are complete ignoramuses. We have psychology instead." D. H. Lawrence

Without knowledge and understanding, it is harder to do well in the real world.

"Each player must accept the cards life deals him or her; but once they are in hand, he or she alone must decide how to play the cards in order to win the game." Voltaire

We cannot play the cards in the game of life when no one teaches us what we need to know to succeed. We must educate ourselves and our young to understand how the forces of life affect us before we have the knowledge it takes to succeed, to go beyond the roadblocks life puts in front of all of us.

CHAPTER 2

ORIGIN OF THE FORCES OF LIFE

"Life in the twentieth century is like a parachute jump. You have to get it right the first time."
Margaret Mead

There are no do-overs.

A newspaper article recently noted an extraordinary fact: *Male children are between two and three times more likely to drown than female children.* What? How on earth could this be? The article had no idea.

Could it be genetic? Do females have inborn, genetically determined, flotation devices that males do not have? Or do males have an "idiot" gene, a gene that makes them take incredibly stupid chances?

THE "IDIOT GENE"

Actually, evolutionary psychologists have proposed that there is a "male idiot gene." They do not call it an "idiot gene," of course; they call it a "risk-taking" gene— that sounds way more "cool" than "idiot gene."

But evolutionary psychology is based on the premise that genes are only preserved if they are an advantage in allowing us to survive long enough to reproduce our own genes. Where is the survival advantage in a gene that kills us off in great numbers when we are that young?

Imagine this; three eight-year-old boys are standing by a swimming hole. One says, "Oh boy, let's go swimming," and jumps in. The second boy says, "The last one in is a sucker!" and jumps in. What is the third boy, who cannot swim, going to do? He is going to jump in and drown. Because he is a guy, and there's that idiot gene.

No, it is something far more important to understand than an idiot gene. If you are a male, growing up in our culture, the one thing you learn before you ever get to school is that if you are afraid to do something, the other boys will taunt you with, "He's chicken! He's a scaredy-cat! He's yella!" and later in life, "Don't be a wuss!" "Grow a pair!"

And, if you are a guy growing up in our culture, you would rather die than have the other guys think you are "chicken."

There are no do-overs.

Even if the other boys do not taunt him, the idea is already there, embedded in his brain, not in his genes; that others will think less of him if he hangs back. The emotion of fear (of what others would think) shapes his brain and controls his behavior.

You can see the same problem throughout society. Fatal traffic accidents, the cost of automobile insurance, death in university fraternity initiations, and admissions to the emergency room at hospitals are all biased against males all the way up to age twenty-five, at least.

The number-one reason for emergency room admissions for male children is sports injuries. We take risks, not because of our genes, but because of our fears of looking chicken, our conditioned fear that others will think less of us if we are afraid, or think more highly of us if we do take risks.

We could probably save half of those boys who drown or die in automobile accidents, or fraternity initiations, if only we took the time to teach them about peer group pressure, of how our minds are programmed by other people's words, of how easily we do stupid things just to fit in. Just telling them, "Watch out for peer pressure" does no good. They need to learn from *repeated examples* in school or at home. Not just once, but again and again.

Every exposure to such examples makes it slightly less likely that the kids will behave the same way if they are put into such a situation.

"If you find yourself on the side of the majority, pause and reflect." Mark Twain

But we do not tell them. We dump our kids into public schools with no thought of what to expect and no understanding of how their minds work. And they sink or swim.

In 2012, thirteen students at the Florida A&M band were indicted for "murder by hazing." I never knew there was such a charge. In the band? I never even heard of hazing in a band. The victim was a respected bandleader who reportedly did not want to go through the hazing but told his parents he felt he had to because everyone else did or he *would not be respected*. What would his peers think?

They made him "run the gauntlet" between two rows of band members who got to hit him as he ran. Some used this as an excuse to hit him as hard as they could. He was beaten to death by his fellow students. It made headlines, in part, because so many of his fellows were charged with the crime, "Murder by hazing."

This is motivated by the fear of "what will other people think" that is at the genesis of what goes on in fraternity initiations every year, and nearly every year, there is at least one death from this reported in the media. We all do stupid things we would never do alone because we fear not being accepted and what others will think. Even if we do not think of this force as "fear", it is still there, hidden in the back of our mind. If we cared enough to teach about the fear of "what will others think" in our schools, we might be able to reduce this problem. But we do not.

The fear of what other people think has its origin in childhood experiences, but its impact is greatest in adolescence. There is an equal and opposite reaction; the positive emotional reaction to what society praises.

THE FORCE OF "WOW!"
Conditioned Emotions

More than this, males believe that other males will look up to them if they are risk-takers because the boys all react with "Wow!" to boys who take chances. "Wow! Did you see that? Tommy ran right out in front of that eighteen-wheeler, he had to hit his brake... that's so cool!" "WOW!" is a tremendously powerful positive conditioned emotion. We learn early that other ten-year-olds admire those who take risks.

All of this is echoed in Hollywood movies—Fast and Furious, Fast and Furious II, III, IV, V, VI, VII, VIII, IX.... Taking chances and being fearless is glorified by the male culture, by Hollywood, on TV, in the military, the police, and in the stories, society has pumped into our brains.

DO GENES DETERMINE OUR FATE?

Where did the idea of a "risk-taking" gene that we hear so much about come from? The Israeli study that first identified this gene said it applied to only 10% of people. No one ever said it only applies to males, yet males clearly have the greatest risk-taking behavior.

When James Olsen and his associates at the Hutchinson Center studied mice with the single copy NeuroD2 risk-taking gene, they found that they did not learn fear as well as normal mice. However, what was not mentioned in the newspaper version of the story is that they also found that these mice did not survive for more than a few weeks. Is this a "risk-taking" gene, or is it just a serious genetic defect in mice that can in no way apply to the average person?

Whether they are talking about an "alcoholism" gene, or a "risk-taking" gene, or a "depression" gene, once the media mentions this to the public, they never bother to put reality into perspective; they never tell the whole story, just enough to tickle your limbic system. The news media goes ape over anything that smacks of hi-tech, such as genetics. They seem to feel that psychology is somehow not real; that it's all in your head.

FEARS EMBEDDED IN OUR MINDS:

Life is not what we were told it is; life is full of pain and problems. Instead of censoring the problems as our schools do, these are the things we need to teach our children, to let them know they can succeed in the face of the worst life has to offer, to teach them what they need to know to survive in the real world.

B. F. Skinner is the man picked by his colleagues as the "greatest living psychologist" back when he was still alive. In a survey carried out among 182 chairs of psychology many years ago, he received more than twice as many "votes" from his fellow psychologists as the next runner-up, who was Carl Rogers. In his autobiography, Skinner writes that, as a child, he learned from the stories told by his grandmother to fear the burning fires of hell if he told a lie and from the warnings of his mother to fear "what other people will think."

Skinner lost his fear of the fires of hell. But one fear that we never entirely lose, a fear that preys on our minds and controls our very thoughts, is the fear of what other people will *think*.

THE ORIGIN OF OUR FEARS

The Book of Lists noted that adults who were asked to choose what they were afraid of from a list said they were *most* afraid of *speaking in public*. Fear of death came in second. People were more afraid of getting up in front of an audience and speaking in public than of death.

Why? Why would adults be afraid of getting up in front of people to speak? Because we learned to fear, in the fire of youth, what other people will *think* of us. To be sure, we have to be concerned about what other people think; our boss, our spouse, our friends. But there is an enormous difference between fear of what others think and a rational understanding of it.

We fear what people will think because we have learned the hard lesson early, from the trauma of our youth. It did not take much. It crept into our minds without our being aware. What happens in first grade when little Bobby drops his tray in the lunchroom? The other kids laugh. What happens if the teacher asks him a question, and he flubs the answer? Everybody laughs. What happens if he trips in the hallway? Everybody laughs.

If we trip and fall in public, before we even get up, we look around to see if anyone noticed us, to see if they might smirk or laugh out loud at us. We do this because we learned early, that others will judge us.

We learn the sting of life just from watching what happens to others. Like learning the language we speak, the idea is subtly implanted in our minds, that others will laugh at us if we make a mistake. If we do something wrong, others will judge us. That experience, beginning so early, takes a toll on us as adults. We learn to fear being judged by others.

As children, we learn other kids make fun of us if we make a mistake. We are held up to public ridicule by others who laugh at us or tease us. However briefly this seems to others who laugh, it is remembered by those who are laughed at, as "the way you made me feel". As adults, we learned from our experience that others cut us down behind our backs or laugh at us if we make a mistake. From the ridicule and abuse of childhood innocents to the office gossip of adults, the same theme is endlessly impressed on our minds. Even as adults, we still worry about what our boss, our spouse, and our fellow employees think about us, the imagined judgment in the mind of others.

The entire Republican Party was terrified of getting a mean tweet from Donald Trump that might make them hated by his followers. They censored any criticism of him, they went along with taking kids away from their parents at the border and putting them in cages, they failed to criticize, even when more than two hundred fifty-thousand Americans had died of Covid-19. 250,000 dead. That is more than all wars since WWII combined. The Korean War killed 58,000, eight years of the Vietnam War killed 56,000, two Gulf Wars killed over 8,000. And if a Republican did tell the truth, what would their voters think?

There are no exceptions. Not adults. Not even Congress. Peer pressure is profound.

As adults, the source of our fears may change, but we may still fear losing our job (what does the boss think?), losing a friend, or losing the respect of others. We are less likely to actually experience fear because we have learned to automatically avoid doing the things that might cause this fear, to censor what may make us disliked.

People may protest, "I'm not afraid!" Yet we are not afraid, because we have learned to avoid doing or saying anything that would make others think less of us. Or, when people intentionally go against the group to "prove" that they are not afraid, they have come to think that proving they are not afraid to go against society is better than letting people think they are afraid of what other people will think; Phobophobia, or a fear of looking afraid.

LIKE AN OAK FROM ITS SEED:
Other People's Opinions Are Implanted In Our Brains

Samuel Clemens, better known as Mark Twain, wrote of its awesome power in his much-censored book *Letters from the Earth*:

> *"Laws are carefully reasoned out and established upon what the lawmakers believe to be a basis of right. But customs are not. Customs are not enacted. They grow gradually up imperceptibly and unconsciously, like an oak from its seed. In the fullness of their strength they can stand up straight in front of a world of argument and reasoning, and yield not an inch. We do not know how or when it became custom for women to wear long hair, we only know that in this country it is custom, and that settles it.*

> *"Maybe it is right, maybe it is wrong--that has nothing to do with the matter; customs do not concern themselves with right or wrong or reason. But they have to be obeyed, one may reason all around them until he is tired, but he must not transgress them, it is sternly forbidden. Women may shave their heads elsewhere, but here they must refrain or take the uncomfortable consequences. Laws are sand, customs are rock. Laws can be evaded and punishment escaped, but an openly transgressed custom brings sure punishment. The penalty may be unfair, unrighteous, illogical and a cruelty; no matter, it will be inflicted just the same . . . Custom is custom; it is built of brass, boiler iron, granite; facts, reasoning, arguments have no more effect upon it than the idle winds have upon Gibraltar."*

What would the "sure punishment" be if you shaved your head? Are you willing to try it to find out? The Hair Police would not slam you against the wall and arrest you for doing so. No moral code would condemn you to the fires of hell. But there is no question about its "sure punishment": The head-turning, the finger-pointing, the whispers, the gossip and laughter behind our backs.

Like the language we speak, other people's ideas and emotions are implanted in our minds. Those words, thoughts, and ideas become the criteria by which we judge our lives, for better or worse.

Look at what happened to Britney Spears when she shaved her head bald. The news media had a field day, like first graders on the playground. "Britney's having a breakdown!" became the mantra, just as if her shaving her head was actually news. She said she did it thinking that if she were ugly, the paparazzi would stop hounding her, as they hounded Princess Diana.

Someone came up with a diagnosis of bipolar for Britney. Really? How would you feel if you were hounded every day by the media? How would you feel if the news media had published a photo of you with your legs spread for a brief second while getting out of your car (appropriately censored with a fuzzy spot by CNN for television)? And then the news media spent days going on and on about you not wearing panties? Would that make you a little bipolar—paranoid, even?

> *"We live on, cheerful, self-confident, until in some rude hour we learn that we do not stand as well as we thought we did, that the image of us is tarnished. Perhaps we do something, quite naturally, that the social order is set against, or perhaps it is the ordinary course of our life that is not so well regarded as we supposed. **At any rate, we find, with a chill of terror, that our self-esteem, self-confidence and hope, being chiefly founded on the opinions of others, go down in a crash.**" Charles Horton Cooley, Italics and bold added.*

Being famous and given the adulation of the masses may give you a great high, but when you fall, you fall from a great height, and the thud is bigger, even if it is all in your mind. Success has a half-life.

Our self-esteem is not determined by reality, or even by what others actually think about us, it is determined by what we *think* others think of us. And our perception of what others think often does not reflect any reality.

No one was more keenly aware of this force than Mark Twain himself. The beloved author of *Tom Sawyer* and *Huckleberry Finn* had written several stories that so powerfully satirized our society and beliefs that his own family begged him not to publish. They feared that the hostility people would feel toward his criticism of our society would turn him from the "most beloved" to the "most hated" American author. Twain himself was more like Jon Stewart of The Daily Show than he ever was as we remember him.

The writings were never published in his lifetime, even though Twain himself owned his own publishing company and could easily have done so. Not until over a quarter of a century after his death did Bernard DeVoto, Twain's biographer, bring these works together to be published. But when Twain's daughter Clara read the manuscript, she was horrified. She refused to consent to publication during her lifetime. Twenty more years went by until finally, it was published in its entirety as the book *Letters From the Earth*.

The worry over "what other people will think" even influences what we are allowed to read in books, what we see on television, what we as teachers tell our students, and what is presented on the news. Each individual in the media strives to give the people what they think they want to hear and avoid what they think they do not want to hear.

Most of us are scarcely aware of the existence of this censorship. We cannot imagine what it is that is censored from us. Every teacher knows we censor certain things from our students, lest it create a controversy, make students feel bad, or appear biased, or simply not be believed.

No one is immune. Any famous man or woman is fair game to the press. The same media that makes a person into a celebrity easily breaks them and blames them for it. Each celebrity comment, every politicians' muse, may be trotted before the glare of public opinion. A scandalous slip here, a misstatement there, tap dancing in a men's restroom, fox hunting in Argentina, fourteen mistresses too many, and the mighty could be savaged on the altar of the god of public outcry. What will people think?

> *"Protection... against the tyranny of the magistrate is not enough; there needs protection against the tyranny of the prevailing opinion and feeling, against the tendency of society to impose, by other means than civil penalties, its own ideas and practices as rules of conduct on those who dissent from them." John Stuart Mill, On Liberty*

How is it possible that over a century after John Stuart Mill wrote those words, we find ourselves unable to teach what Mill knew to our own children?

SOCIETY IMPOSES "WHAT WILL OTHER PEOPLE THINK?"

> *"Society is an interweaving and interworking of our mental selves. I imagine your mind and especially what your mind thinks about my mind and what my mind thinks about what your mind thinks about my mind. I dress my mind before you and expect that you will dress yours before mine. Whoever cannot or will not perform these feats is not properly in the game." Charles Horton Cooley*

"Many a youth who is demure around his parents and teachers will strut and swear like a pirate among his tough young friends," said psychologist William James. Why do we behave differently with our friends than with our parents, employees, or supervisors?

There are things we dare not say to our parents, supervisors, and fellow employees. Why?

What is it that keeps you from singing out loud to yourself in a bus full of strangers as you might sing loud by yourself driving alone in your car?

William James noted that we change even the tone of our voice, not to mention the type of language we may use when we are in the presence of a minister or clergyman. Why?

Candid Camera once did a show where cast members waited near an elevator until an unsuspecting man came along. The man went into the elevator and turned around to face the door. Immediately, one of the cast members went into the elevator and faced the rear instead of turning around to face the front. Then a second cast member did the same. The first man glanced up at both furtively. Then a third and a fourth cast member came in to face the rear of the elevator.

Looking around sheepishly, the first man then turned to face the rear, the way the majority faced. Not a word was spoken. No one made fun of him. Yet we all learn, from those harried experiences of youth, to follow the lead of the majority, lest we be tarred with the brush of being "stupid" or laughed at for not knowing what everyone else seems to know.

Not only do we shape our behavior to the tune of what others will think, but we also imagine, from what we have already learned, what other people will think. No formal recognition of the problem is needed.

What are we afraid of? Would they beat you severely about the face and head with an umbrella for facing the wrong way in an elevator? Or what if you were alone in your car singing out loud to the music on the radio and stopped at a stoplight, and another driver looked at you while you were singing out loud to the music? Would the music police arrest you? Not likely.

The worst that other people are likely to do is glare at you with a jaundiced eye, what George Herbert Mead called *"the sidewise glance,"* the scowl of disapproval that leads to the imagined judgment in others' minds. Or, even worse, they might laugh at us.

There is a quaint commercial about this fear. A man is driving along in his car, singing and rocking along with the beat on the radio. He pulls up to a stop sign, still singing. Another car pulls up alongside. The man looks over, only to see the man in the other car looking at him. He immediately straightens up, stops singing, and looks straight ahead.

What will other people think?

We do not stop to analyze our response; it all seems so natural, so obviously correct. Thought is not required. We see the fear in children's faces, but by the time we are adults, the fear is replaced by a simple perception—we are aware of what other people think, so we avoid doing whatever might trigger the fear itself.

We learned early on to adapt our behavior to the idea of what others will think. We learn to avoid the fear, because we do not violate the unwritten laws.

The unwritten laws are so deeply rooted in our mind that they shape every action and thought, including our beliefs, interaction with others, sexual behavior, personal habits, and politician's opinions; even the stories we are allowed to hear from the news media all bow to the force of public opinion. It exercises its control over our lives without our ever being aware of its force.

"The highest possible stage in moral culture is when we recognize that we ought to control our own thoughts." Charles Darwin

A student once said to me, "I see, psychology is just common sense." No. It is only common sense *after* it is pointed out. Even then, few understand the enormity of its influence on our mind.

THE BEGINNINGS OF FEAR

We have all seen it happen to others. We know what will happen to us if we do not measure up to other people's expectations—they will judge us.

Gestalt psychologist Fritz Perls famously said, *"I am I and you are you. I am not in this world to live up to your expectations and you are not in this world to live up to mine."* It is a great saying, but it does not change reality. Before we can change, we must understand how society produces the expectation that we must live up to the expectations of others, and how to change that in our own mind to something more realistic.

Sometimes, these fears are useful. They may make us stop and think before we act, to consider how our actions affect others. It is important to be considerate of others. Yet, often, these fears make us afraid to act. They inhibit us from going beyond the limits that others impose.

In a sense, we never escape the power of "what other people will think" because we must be concerned about what our boss thinks of us, what our friends think of us, what our spouse thinks. These are what George Herbert Mead called the *"significant others"* around which our universe revolves. Hopefully, it will be a rational understanding, rather than an actual fear. If we want to do well in life, we must consider the feelings of others. Life is like a Rubik's Cube.

Are adults immune from peer group pressure? Hardly. A trip through YouTube finds a striking parody of this, echoing what we see in fraternity initiations. In one Japanese game show, male contestants have to recite a paragraph within a time limit without making a mistake. Five males are required to stand spread eagle on special platforms on a stage. If they make a mistake, the host blows a whistle, and a wooden plank springs up from below them and whacks them in the gonads.

One after another, the men scream and clutch their privates. Even contestants and viewers who do not get hit, cringe every time it happens.

The last contestant repeatedly tries to leave before it happens to him, but every time, he was told by the host to be a man, to get back up there. No law would make him go through with this. No one could legally stop him from leaving. Yet, just like college students in a fraternity initiation, he blindly obeyed; something else we learned in school, the military, and society. He got whacked in the gonads.

It is a good video to show when talking to your kids about peer group pressure, at least before they go to college or join a fraternity.

FREE WILL?
Of Worms and Rocks

Psychologists may debate in-depth over whether humans have "free will" or if our behavior is "determined" by forces in our environment. Do we have no more free-will than Pavlov's dogs?

I suspect it is largely a matter of degree. A worm has more "free will" than a rock. A dog has more "free will" than a worm. A human, arguably, has more "free will" than a dog.

Yet, if we are unaware of the forces in the environment that determine our thoughts, emotions, and behavior, do we have any more "free will" than Pavlov's salivating dog? Do we all just emote when someone rings our bell?

Learning to understand the forces in life that shape and determine our lives gives us the potential to gain greater control over our own minds. If we fail to learn what we need to know about the forces in life that affect our minds, then we would just as well be salivating to the bells that others ring to control us. Free will is not a fact; it is only a potential. It will vary in degree.

Without a thorough understanding of how the forces in our lives affect our minds, of how others shape our minds, we cannot protect ourselves or our children from life's problems.

THE BIOLOGICAL BASIS OF FEAR

"If a man picks up a cat by the tail it will teach him a lesson he cannot learn in any other way." Mark Twain

In the entire of our desperate history on this earth, our ancestors survived because our brain's biology had mechanisms designed to help keep us safe. The amygdala, part of our emotional limbic system, is the center of anger and fear. Biologists call it the *Fight or Flight* response. When this center is triggered, adrenalin, epinephrine, and cortisol shoot into our bloodstream. Our heart jumps, and it prepares us to engage in a massive struggle for existence, or it makes us quickly drop the cat.

In our past, this was an advantage. What we feared, occupied our minds and forced us to pay attention to the problem. If we worried about starving during the winter, that emotion of worry forced us to plant more crops, harvest more game, and work harder in the anticipation that we might starve. If we worried about being attacked by our neighbors, that emotion motivated us to build our walls higher and make more arms for defense out of the anticipation that we might be attacked.

Yet today, we rarely have to worry about starving or our neighbors attacking us. What do we worry about? Trivia. We lay awake at night worrying about whether someone likes us, if we will have a date for the weekend, if our boss might give us a bad recommendation, about the loss of love, or the respect of others. These are not life-threatening worries.

Often, there is nothing we can do about what we worry about. Yet, the brain is not easily capable of distinguishing between life-threatening worry and worries about trivia.

The brain reacts the same to trivial worries as it does to life and death issues.

"The human race is a race of cowards; and I am not only marching in that procession but carrying a banner." Mark Twain

When neurosurgeon Wilder Penfield electrically stimulated conscious patients' temporal cortexes, he found far more negative emotions were elicited than positive. This is likely the outcome of a culture that uses fear and anger far more than positive emotions. That begins at an early age.

WHAT WOULD MAKE YOUR HEART RATE GO FROM 76 TO 143 IN SECONDS?

National Geographic tells a remarkable story of fear. They have one woman who is terribly afraid of feathers. How could anyone possibly be afraid of a feather? On the show, they connect her to a heart monitor to measure her fear. As soon as they show her a feather, the sight triggers an electrical impulse to her amygdala, adrenalin pumps into her system, and her heart rate jumps from 76 to 143 beats per minute, in only seconds—no one can fake that; we have no such conscious control over our autonomic nervous system.

How could this happen? Her grandmother says that this began when she was in a room with a trapped bird as a child. If you have ever seen a bird trapped in a room, you know it will panic and flap around, feathers flying and banging into windows trying to escape. Children have no understanding of what is happening; they may be frightened. Her grandmother tells her that, *"...you were terrified."* Even though she does not remember the incident, this was probably the start of the phobia.

"Once bitten, twice shy." Unknown sage.

THE 'SNOWBALL FROM HELL' SYNDROME:
The Brain Forces us to Selectively Focus on Our Fears

But that is not the whole story. What would have happened in first or second grade, when the other kids found out she was terrified of feathers? Little boys would torment her. "I've got a feather" and try to put it on her, frightening her even more. This would have likely made the fear even worse than before.

Once a fear is learned, the brain automatically focuses our attention on the thing we fear. As she got older, she would no doubt have heard of stories like Alfred Hitchcock's movie, *The Birds,* where an entire town was taken over by birds that pecked people's eyes out. Many more movies have since come out, such as *Kaw* and *Resident Evil: Apocalypse,* showing birds attacking people, complete with panicked reactions from the actors.

As the snowball rolls downhill, it picks up more and more snow. Likewise, the first experiences embedded in our brains pick up more and more similar experiences or ideas as we go through life. Not just our fears, but our emotional reactions in politics and religion pick up more and more ideas that support the first ideas embedded in our brains, and cause us to ignore the evidence against our beliefs.

Is it hard to believe that anyone could be afraid of a feather? For thousands of years, our entire culture was terrified of witches and demons. Being afraid of a feather is not so different. It took a long time to desensitize our entire culture so we would no longer be afraid of "things that go bump in the night." How could an entire culture be terrified of things that do not even exist? How could this go on for thousands of years?

WITCHES AND DEMONS WERE FEARED FOR CENTURIES

Feathers are real, witches and demons are not, yet the fear felt by the audience at Salem as the girls began to shriek *"witch, witch"* sent chills up their spine. Adrenalin, epinephrine and cortisol shot into their blood. Their hearts jumped. That was all it took to convince adults their fear was real. The response of their own body convinced them witches were real.

Today, the same evidence is used to convince people their problems are caused by their biology, their DNA.

Don't let the fears others put in your brain, control your mind.

The emotional reactions of others, like the girls shrieking *"witch, witch"* at Salem, are all it takes to embed fears in our minds.

These were not witches; they were our grandmothers. Witch hunts had nothing to do with ergot poisoning or "real" witches—it had everything to do with words, thoughts, and ideas that triggered emotional reactions in the brain. The stories we hear about Salem were just a pimple on the behind at the end of a terror that swept through Europe for thousands of years. Based on court records and news accounts, more than 30,000 of our grandmothers were put to death as witches in the 16th century alone; the first century we have records for, thanks to the printing press.

We may no longer fear witches, yet every generation invents new "witches" that frighten us; from "weapons of mass destruction" in Iraq, which turned out not to exist, to fear of Indians, which led us to kill or imprison them in concentration camps in deserts and badlands, to labeling every problem as a "mental illness." Words, associated with an emotion, are all it takes.

Words determine who we love or who we burn.

People think it is absurd to be afraid of a feather, yet is it any more absurd than to be afraid of "what other people will think"? The difference is that few of us learn to be afraid of feathers; all of us learn to be afraid of "what other people think".

After a while, the *fear* of what other people think goes away, because we learn never to do or say what we know would trigger that fear. Yet the behavior remains.

The woman who was afraid of feathers had earlier gone to her doctor to ask if someone could help her overcome her fear. *"He looked at me as though I was crazy,"* she said. This is not something we always teach in medical school. If doctors only think in terms of medicine, you are likely to end up on Valium or Ativan or Buspar. That would not have helped at Salem.

When we get older, we may no longer experience an actual fear associated with a word or idea, yet we still react with judgments of others based on what emotions have previously been associated with words, thoughts, or ideas. Those early experiences determine our reactions to others, our judgments of ourselves, and our political and religious opinions that we hold with such fervor that we are unable to change or even consider the evidence that other ideas might be right.

The snowball effect of learning comes from the brain, automatically forcing us to pay attention to emotional ideas already deeply ingrained in our brain. One of our culture's greatest fears was once of the number 13, as in "Friday the 13th," the movie.

People who were already afraid of the number 13 were the first to notice incidents like the Apollo 13 ill-fated lunar landing attempt that almost ended in disaster. Those who were not afraid of the number 13 were unlikely to even notice that it was numbered 13.

Those who were afraid of the number would say, "See. That proves that number 13 is unlucky!" Even though the times we lost astronauts were not numbered 13 at all, they were less likely to notice that.

This "snowball" example had a far more profound effect on those with this fear in their brain than on the rest of us, who never had such a fear. The brain selectively focuses our attention on whatever ideas are already embedded there— whether it is the fear of feathers, witches, Indians, weapons of mass destruction in Iraq, or political and religious fears others use to control us.

You find the same effect in our politics and our religions, where the brain quickly attends to information that agrees with the ideas first embedded in our brain, and easily ignores evidence that disagrees with our ideas. The conflict over the election of 2020 is sad evidence of just how devoted the brain is to whatever ideas, even nonsense, that is first put in our brains by others.

How do you get rid of a fear? She went to see a cognitive-behavioral psychologist who used the techniques pioneered first by Mary Cover Jones and later made into a cognitive therapy by Albert Bandura and others. Counter conditioning, desensitization, and control are the important methods.

He first had to convince her to even look at a very small feather. She could not. It took him twenty minutes to get her to look at a feather at a distance. When he picked up the feather, she trembled.

"My holding it makes it worse?" he asked.

"Yes, because you could put it on me," she replied.

That statement may be diagnostically important; it strongly suggests that other kids in school had done just that, threatening to put a real or imaginary feather on her just to frighten her, which would have made the fear even worse.

GETTING CONTROL OVER OUR MINDS

He began with the feather some distance away from her. He placed it on a platter and gradually moved the feather closer. He gave her **control** over how he moved the platter. She could tell him to move it closer, move it back, or stop. This combined *gradual desensitization* with giving her *control*.

By giving her control over the feather and having her say when to move it closer, stop, or move it back, it was possible to gradually desensitize her to the fear, leading to the extinction of fear, without terrifying her. It took 45 minutes, but she was able to hold the feather and stroke her hand with it.

When told it took 45 minutes to eliminate her fear, her response was, *"Wow, forty-five minutes after thirty-seven years."*

Do not bet, however, that fears of "what other people will think" that is reinforced over a lifetime, including fears of our parents, friends, and bosses, can be easily extinguished in 45 minutes. It is harder to gain control over what others think than over a feather, but it can be done. Understanding how this has affected the minds of all of us, is the beginnings of being able to control our own minds.

We spend much of our lives trying to gain the skills necessary to have more control over our life. Control begins with understanding while experience gives us the skills, and the more control we have over our own mind, the better off we will be.

There are over 700 named phobias; from fear of cotton to fear of Santa; fear of feathers is only one. The fear of bugs is probably the most common in our society. Yet, the most commonly treated phobia is our social one, triggered by the "thought" of what other people will think.

Any stimulus -- Any Emotion.

SOCIETY PROGRAMS THE EMOTIONS IN OUR BRAIN:
Conditioning of the Peer Group, by the Peer Group

The programs that others imbed in our brains become the blueprint for our perception of reality. It is amazing how little it takes. "Big boys don't cry!" was a term most of us heard from an early age. Sadker carried out studies of how first-grade teachers, all females, treat boys and girls differently. If a girl hurt herself and went crying to the teacher, the teacher would sympathize, "Oh, you poor thing," and bandage up the hurt. If a little boy went to the teacher, she might bandage it up and say, "There now, that didn't hurt, did it?" the implication being that "big boys don't cry."

This persists into adult life. In a study, college males volunteered for a psychology test where they were told it was to be a test of pain threshold. They would be wired up and started out with a very low level of mild shock. Each time, it would increase a little more. When it became painful, they were to tell the experimenter, and that would stop the experiment.

The first experimenter was an older man who wired them up and gradually increased the shock level. The males would quickly reach the painful level, and the experimenter would stop the experiment. He would then excuse himself, ask the male to wait just a bit, and shortly, a new experimenter would come in—a young female their own age. She repeated the experiment, and guess what? The males took way more shock from her than they took from the older man.

Why? Males do not care so much what an older adult thinks of them, but someone their own age... they did not want to look weak. Even if the second experimenter was a male their own age, they still took far more shocks for the young man than for the older.

This persists for much of life. Any psychologist or psychiatrist will tell you that 70% of their patients are female—males do not want to admit to having problems. We are supposed to be able to suck it up, to take it, to be tough guys, to handle our own problems.

When we were talking about this, a female student in class said that when her own boyfriend complained about something, she told him, "Buck up, be a man about it." So, we get hit from all sides—the girls as well as the guys.

Of course, there may be good reasons not to complain about every little thing.

When women have problems, say an argument with their boyfriend or spouse, what do they do? They call their mom, they call their sister, they call their best friend, until they get others to agree that he is being just as big a jerk as she thinks he is.

When a man has an argument with his girlfriend or spouse what does he do? Does he tell the other men at work about it? No. He doesn't want to look like a wuss who can't handle his own problems. He might say, "Aw, the old bitch was on my case." But he will not talk about it. Instead, males in anguish will *"slip on down to the Oasis, where the whisky drowns and the beer chases..."* Garth Brooks. In the male subculture, it is more acceptable to get stupid drunk than it is to talk about your problems.

There is even an old joke about this. A woman is awoken in the middle of the night by noises coming from the kitchen. The noises go, Bop...Ding...Bop...Ding...Bop.

She goes into the kitchen to see what is happening. Her husband is standing in the middle of the floor with a fly swatter.

"What are you doing?" she asks.

"I'm killing flies," he says.

"Did you get any?" she asks.

"Yep. I got five. Three males and two females"

"How do you know if a fly is a male or a female?" she asks doubtfully.

"Easy," he said. "Three were on a beer can. Two were on the phone."

It goes to our culture's cliché over what the difference is between men and women. Men are drinkers. Women are talkers. Of course, not all men are drinkers and not all women are talkers, but clearly, everyone recognizes this cliché.

POST TRAUMATIC STRESS DISORDER:

The real consequence can be seen in problems with males and Post Traumatic Stress Disorder, or PTSD. A recent government study says that 39% of soldiers in combat in the Gulf War will suffer from either PTSD or Major Depressive Disorder, often years after the war was over for them. Males are simply ashamed to come in for help, even when they suffer from severe symptoms, because they see it as an admission of weakness to admit they cannot control their own mind.

So, males in anguish will "*slip on down to the Oasis, where the whisky drowns and the beer chases...*" Garth Brooks. In the male subculture, it is more acceptable to get blind drunk than it is to talk about your problems.

Yet, the number one cause of PTSD is not weakness, but the amount of blood and death the individual is exposed to. That sucks the feeling of security right out of them. We know this because the ones who suffer most from PTSD are the "grunts" on the ground—a term they call themselves.

Those who fly helicopters or planes that shoot hellfire missiles that may kill dozens never see the end result of what they do up close. Only those who see the effect, the blood and death that sucks that feeling of security out of you, or can feel the guilt of having killed others, suffer from PTSD.

APPROACH-AVOIDANCE CONFLICT

One of the most common problems we all face is the Approach-Avoid conflict. This is a mental conflict that occurs when the same stimulus triggers both a positive emotion (Approach) and negative emotion (Avoid) at the same time.

We see this constantly. We may want another piece of pie (Approach), but then we have to worry about gaining weight and "what other people will think" (Avoid). It is all about words, associated with positive or negative emotions.

We may want to have sex (Approach). But then we have to worry about "what other people will think" and pregnancy, syphilis, gonorrhea, chlamydia, hepatitis B, AIDS, human papillomavirus, genital warts...(Avoid).

Usually, we only worry about that much later.

Emotions in conflict dictate much of human life, but rarely is that conflict of emotion as great as in combat. In a brilliant film with Candy Crowley, done by CNN on Post Traumatic Stress Disorder, we see a dramatic conflict between our social and religious values "Thou shalt not kill" (Avoid) and the way we train our soldiers before going into battle to be heroic and kill the enemy (Approach).

In WWII, the military learned that most of our soldiers never fired a rifle at the enemy (they did not want to kill). So, the military set out to remedy that problem.

They trained soldiers on pop-up human-like targets instead of standard target shooting. To qualify, they had to shoot as soon as a target would pop up, without thinking.

As the drill sergeant described it, *"It becomes an unthinking, conditioned reflex. Stimulus-Response, Stimulus-Response,"* See it—Shoot it. See it—Shoot it. That solved the problem of eliminating the thought of killing, by replacing thoughts with an "unthinking" automatic response.

One soldier from the Vietnam war described how, when some men failed to qualify on the rifle range, the sergeant gave a speech, *"These men have failed their country."* Then he made the men lie down on the ground, and the other soldiers run over them again and again. Eventually, they could no longer control their feet and were stepping on their face. The training of the military shames us if we fail to do our duty and kill the enemy.

The soldier with PTSD turns away and cries after describing how he killed a Viet Cong. *"There is a lot of feeling about it later, not at the time, though, not at the time."*

He is describing a traumatic conflict inside one's own mind over the Approach (See it—Shoot it) success in training, and the Avoid (of killing another human being).

A second soldier from the Gulf War, describes machinegunning a group of Iraqi soldiers who were trying to escape a burning truck. He said he had a moment of exhilaration (over doing what he was trained to do), followed almost immediately by a feeling of remorse (over killing others). He later wrote a letter to his girlfriend back home about the night, but tore it up and never sent it because he was afraid she would think badly of him.

He was doing what he was trained to do.

Perhaps the great irony of PTSD is that it is often the best people, the most thoughtful and sensitive people, are the ones hurt the most by their experiences.

As long as one can rationalize killing, "kill or be killed", there may be no conflict for some. Yet most kills are not "kill or be killed". But when, usually later, one thinks that the person they killed was just a grunt, like me, who had a wife, family, and kids who loved him—that *thought* creates an approach-avoidance conflict in the mind that makes it hard to bear.

Killing is not necessary for PTSD to occur. Just seeing the blood and death of so many others sucks the feeling of security right out of you. If our childhood belief in our own safety is broken, that alone is enough to create a sense of loss, the loss of our confidence in life.

MEMORIES FLIT THROUH THE BRAIN,
triggering emotions that create anxiety or depression

Basic to the problems of anxiety or depression are the memories that repeatedly flit through the mind of the individual. Past failures, negative experiences, put-downs, mean tweets, all of the things that have happened in the past may be triggered by experiences happening today. These may race through the brain, tripped off by a bad thought, a similar experience, or even just a feeling of failure. Sometimes there is no apparent reason for these thoughts flooding into our mind.

These memories may not appear at all when one is engaged in work at a job, or in talking to others. Activities keep the thoughts our of our mind. But at night, or when we are alone, they may pop back into the mind. That is the reason for the best advice for getting over the death of a loved one, is not to take time off, but to get back to work, to immerse the mind in experiences that keep us involved, that keep us busy, that provide new experiences that take our attention off of the memories that create the problems of life. When we are alone, we worry.

The brief memories that create the problems of life are a result of the frequent reflection on all of the bits and pieces of memory that produce anxiety or depression. Often there is no emotion associated with the behavior at the time but it occurs later, when the mind is free to reflect on the past. Night or alone times may be the worst. PTSD is only one such example.

Even in cases of PTSD, showing the CNN film where other soldiers have been through this, felt badly, and yet were good people, may be enough to help them get through much of their pain or guilt. I know of no study that has attempted to use the CNN film for this, although it should be done, but group therapy, with others who have been through similar experiences, is an attempt to do something quite similar by talking out their feelings. This again, can provide an opportunity for counter conditioning and desensitization to occur.

CONDITIONING OF THE PEER GROUP BEGINS EARLY

I remember when I was just a boy, maybe six years old; other boys called me a "mama's boy" because I kissed my mother goodbye. I did not know what a "mama's boy" was, but I could tell from the tone of voice that it must be something truly despicable. After that, I never wanted to kiss my mother goodbye again; I never told her why. I was made ashamed of kissing my mother by the other boys' words.

Emotional conditioning of the peer group, by the peer group, happens when we are too young to know any better, when we have no ability to defend ourselves from others' emotional words. Our schools have failed to teach us what we need to know to understand what is happening to us.

Today that seems absurd, yet when we are young, we have no control over our own minds; our brains uncritically adopt whatever society feeds into them, even what our fellow morons in school feed into them. As adults, we often cannot see how the power of emotional conditioning by the peer group still controls the very emotions in our brain.

We learn our emotions the same way we learn the language we speak, by the words and actions and tone of voice of others.

I often hear from mothers that they had a good relationship with their sons when it suddenly changed without any apparent reason. There is always a reason. We may never know what the reason is, when it comes from something we never see.

We are often unaware of how our minds can be changed by the emotional conditioning of the peer group. We dump our kids into the school system with no idea of what they will run into. The older kids condition the younger ones, and we all condition each other. Words are all it takes, and the *emotion associated* with those words.

Parents are often unaware of anything that goes on in school, and we never prepare our kids for what to expect. Even simple experiences can leave their mark. As children, we go through many such experiences in growing up, and we never tell our parents. We are often quite unaware of how words, tagged with emotion, shape our lives.

Fears develop in normal people all the time. Children have no fear of dirt; they love playing in it. But at one point, most children do become afraid of dirt, usually after hearing about germs and how they cause disease from the school nurse in a lecture about washing your hands. So, it is common to find children developing a phobia about dirt at some point in their lives, including compulsive hand washing.

But children cannot avoid dirt—they are children. So, they often gradually desensitize themselves just by getting dirty and not being able to wash. Occasionally, you will find this persisting into adults, as with Howie Mandel, who calls himself a "germophobe" and will not shake hands with others. One has to wonder how he has sex. Perhaps a full-body condom?

The same phobia was found in Michael Jackson. At one point, he had to wear gloves and a surgical mask, although he seemed to have gotten over it. You see this in the great adventurer, pilot, and billionaire, Howard Hughes. At one point, Hughes would not even sit in the same room with another person; instead, you had to talk to him over a two-way camera-television setup.

If a new pandemic were to hit America, then you would see most of us converted overnight into "germophobes," wearing face masks and rubber gloves, just as they did in the Great Flu Pandemic of 1918 that left an estimated 50 million dead worldwide.

Oooops. It just did. Yet at least a third of Americans refuse to wear a mask. What happened? Some worry about what others would think of them if they gave in and wore a mask. It became a political statement. Some do not believe the scientific evidence. Some do not care if they infect others. People have different experiences, different emotional conditioning.

People are complicated.

WE DO NOT ALWAYS NEED A PSYCHOLOGIST:
Treating Ourselves

All of us go through fears, and we desensitize ourselves. We do not need to pay a psychologist $150 per hour for most problems. Suppose you are driving in your car, and someone pulls in front of you; you hit the brakes. Your car skids. You stop. Even if you don't have an accident, adrenaline may shoot into your bloodstream, your heart jumps, and fear flits through your brain.

Even if we only get a ticket, what happens next? For the next two weeks, we drive reeeeeeeally carefully. Then, after a couple of weeks, we desensitize ourselves and go back to driving like Vin Diesel.

But we have to get back into our car and drive to get to work, the store, school, and for many other reasons. We cannot afford not to. We desensitize ourselves. A few people become agoraphobic, and they refuse to leave their homes. These are mostly people who can afford to have someone else do things for them or have food delivered. They sit in their homes and watch the horror of daily television, with stories of murder, rape, and kidnappings, and become increasingly afraid to go out. This creates a snowball effect, intensifying the fear.

ANTICIPATORY ANXIETY

"Fear of danger is ten thousand times more terrifying than danger itself... and we find the burden of anxiety greater... than the evil which we are anxious about."
Daniel Defoe

Anxiety comes from the brain anticipating that something bad will happen. Words, thoughts, and ideas that others put into our brains come to control our thoughts and behavior. We have little awareness of where they came from, how they affect us, or how to escape its control over our minds. It all happens with the same subtle force of learning that embeds our language into our brains; without our awareness.

All the emotions embedded in our mind as children become the forces that lead to success or failure in life. In psychology, we may give clients a 25-cent diagnosis of "anxiety disorder," "clinical depression," or "personality disorder," or even "genius", but the words explain nothing.

Psychologists need to think of diagnoses in terms of the experiences that create the problems, not as labels for symptoms that mean nothing. When we can rewrite the DSM-5 diagnoses in terms of the causes of problems, instead of a list of symptoms, we will finally accomplish a major coup in the history of psychology.

Other emotions lead to success in life, and we marvel at their accomplishments, but labeling them as special or genius does nothing to explain what emotions and experiences led to their success. The rest of this book will look into how we deal with problems that others create, how we can stay sane in an insane world, and what makes for success in life from those who have gone on to be a success.

Perhaps the most profound comment on the nature of fear and the problems of life comes from a book by Lara Jefferson, *These are My Sisters*, who, after coming away from her time in a mental hospital, made one of the most important comments on fear:

"At least I have learned this: Nothing is ever so bad when it is actually happening to us as when we are dreading, fearing, anticipating it. It is the fear we build in our mind that gives a thing the power to cause us greater pain."

Like Pavlov's salivating dogs, the bell ringing automatically triggers saliva in *anticipation* of food; no thought is required. In this case, our brain's thoughts are the bell that triggers that twinge of anxiety in *anticipation* that something bad will happen, or a flash of memory of bad things that happened in the past. And those thoughts that trigger this *anticipatory anxiety* can be even more painful than if the worst does happen.

PUT LIFE IN PERSPECTIVE

Learn to put reality into perspective; most of the worries of life never happen. Yet, the brain is not easily capable of recognizing the difference between life-threatening fears and the trivial ones that control our minds.

Think about this; what was it you most worried about a year ago or five years ago? The chances are you are not worried about that today. What you are worrying about today probably will not be what you will worry about a year from now. Practice thinking about this every time you find yourself worrying about something. That helps put reality into perspective.

"I've had a lot of worries in my life, most of which never happened." Mark Twain

CHAPTER 3

STAYING SANE IN AN INSANE WORLD

"...those who were seen dancing were thought to be insane by those who could not hear the music. " **FRIEDRICH NIETZSCHE**

In a remote African range, swept by sun and bright with hope, Elspeth Huxley wrote about an extraordinary event in *Flame Trees of Thika,* which PBS made into a TV series. Terror suddenly struck her friend and father's trusted worker. Overnight, it sucked the enthusiasm from his life and destroyed his will to live. She describes him as *"something of a dandy,"* smiling and happy in his picture. Unlike the happy man he had been, he now stared into space, barely speaking. He left his work and took himself to his home, where he stayed alone in bed in his room, refusing to eat or care for himself.

When Huxley's father went to see him, he found a man without hope, resigned to his fate. *"He lay immobile in his dark hut, fading away like an old photograph. His ribs stuck out, his pulse was very feeble, and he scarcely seemed to breathe."*

They took him to the hospital in town where the best of Western medicine could not help him. They sent him home, *"...his legs like sticks... hunched like a chicken..."* with a note that said they could do nothing for him; that he was resigned to his death.

Why? What could take a man so full of life and drain his mind of energy overnight? The symptoms described by Huxley are common in depression. No virus or bacterial infection caused it; no fevered hormones altered his emotions. No serotonin depletion sapped his pleasure in life; no blowfish toxins had made him a zombie. He was in good health. He had been hexed by his own mind, dying from ideas others had embedded in his mind.

In the backlands of Africa or Haiti, the victim would have seen the warnings; perhaps the headless chicken, blooded feathers made into human form, or felt the prick of the eighteenth pin in the voodoo doll. The priest had cast the bones. Black magic had overwhelmed his mind. The man was going to die.

He knew he was going to die because he had heard many tales from his village elders of others who had died of black magic. The storytellers of his tribe are the Netflix of his life. The stories became part of his mind, dominating his thoughts. He believed it. So did everyone else—his village shunned him.

THOUGHTS AND IDEAS EMBEDDED IN OUR MINDS

Ideas planted in his mind were put there by his own people... stories told for generations by his family, friends, and religious leaders. Thoughts are all it took to produce this terror.

How could simple thoughts come to have such power?

Think about this. From the time you were old enough to watch television, you have grown up in a world in which you have heard stories about cancer. Movies such as *Love Story* tell of its fate; an apparently healthy person discovered that she was going to die of seemingly innocent symptoms. You may have read the *"Ten Early Warning Signs of Cancer"* in Readers Digest.

You have heard of the symptoms. What if you discovered a lump, growth, small tumor, or skin discoloration? You have seen the warnings. What would you think? What feelings would cross your mind?

Do you find a doctor to make a diagnosis? Do you take a chance? Do you really want to know? Merely the *thought* that you might have cancer can generate powerful emotions.

Suppose you go to a doctor who takes a sample for a biopsy and tells you to come back in a few days. How would you feel while awaiting the verdict? Eighty percent of all biopsies are not malignant, not cancer. But that is not likely what you are thinking as you await the results.

And if the doctor told you do have cancer, that it has spread, it has metastasized, how would you react? What emotions would surge through your mind?

Betty Rollin, an award-winning TV journalist who wrote about her experience with breast cancer, put it well in the title of her book, *First, You Cry.*

Fear of death may plunder the enthusiasm for living. Anger and resentment are often felt against friends and relatives who they feel cannot understand the fears they face. Ordinary tasks become difficult. The simple pleasures of life now give no joy.

A belief, a single thought, can change you from being a happy, outgoing person to the opposite extreme in seconds. No physical change has occurred. You would feel no physical pain. This is not caused by any mutant genetic code that suddenly engulfs the mind; no errant chemicals in the brain unleashed demons in the dark side of your mind, no ergot poisoning pricks the mind, no serotonin imbalance will be found in your brain, and no psychic conflicts or early trauma triggered its force. It is caused by something more profound:

"...after Dwayne was carted off to a lunatic asylum in a canvas camisole, Trout became a fanatic on the importance of ideas as causes and cures for disease... But nobody would listen to him. He was a dirty old man in the wilderness, crying out among the trees and underbrush; 'Ideas or the lack of them cause disease'" **KURT VONNEGUT VIA KILGORE TROUT**

IDEAS, TAGGED WITH EMOTION, EMBEDDED IN THE MIND.

Reality may be quite different. The doctor could be wrong. The doctor might have read the wrong biopsy. The laboratory may have made a mistake in samples. Yet, the emotions that envelope your mind would be the same *whether or not you had cancer* just because of the emotions in your mind that the stimulus of the idea of "cancer" triggered.

Suppose your doctor called days later and told you that he had read the wrong biopsy, that it was not malignant, that you do not have cancer. Would you not feel an immediate emotional change, a surge of relief?

Our culture's stories become embedded in our minds. The emotions in the stories become our emotions. It does not matter whether or not something is true; the first ideas implanted in our mind are the template from which we will judge all else.

A recent news story told of a woman in the northeast who was diagnosed with cancer. She had a double mastectomy. When the Board of Tissues examined the tissue, they found no evidence of cancer. She had believed it for months, only to find that the doctor who examined the original biopsy had made a mistake; she had never had cancer.

Anthropologists have found that victims of voodoo rarely die. The stories of voodoo death, so firmly imprinted on the victim's mind, are largely a myth. If a happy, healthy man suddenly dies of an unknown cause, they do not think, as we might, that they died of a stroke, heart disease, or infection—they immediately think he has died of black magic. The stories spread. The myths take on a power of their own, the power to control the emotions in our brain.

The man in Elspeth Huxley's story had a happy ending. The chief of his village found a witch doctor who took off the curse by *words*, telling him the curse had been removed, and that restored him to normal. All it took were counter words or counter conditioning. He recovered after a month of depression and went back to being his normal self. It is all done with words tagged with emotions, thoughts in the brain, a conditioned emotional response, the association of an idea, a word, with an emotion.

"Reality" is determined by the stories and emotions alive in people's minds. Our culture's stories have enormous force. Their texture is a part of our minds. Their emotions become our emotions.

> *"REALITY IS THE #1 CAUSE OF INSANITY AMONG THOSE WHO ARE IN CONTACT WITH IT."* EDGAR ALLAN POE

Our mind has its origins in our experiences. Unless we understand the forces that shape our minds, we will be prisoners of our experiences. We must discover what can be done to prevent psychological problems in our children and ourselves. More than this, we must learn how to control the forces that affect our minds and determine the course of our lives.

> *"The world we see that seems so insane is the result of a belief system that is not working. To perceive the world differently, we must be willing to change our belief system, let the past slip away, expand our sense of now, and dissolve the fear in our minds."*

WILLIAM JAMES (OFTEN CALLED THE "FATHER" OF AMERICAN PSYCHOLOGY)

For thousands of years, our doctors opened people's veins and bled them to make them well. For decades doctors refused to believe the "germ" theory of disease; that tiny, microscopic organisms could kill a big adult.

Psychology and psychiatry today are at that same tipping point. The ideas we learned from our textbooks do not deal with the serious problems of life. Instead of trying to understand the more basic cause of the problems of life, in our emotional conditioning and skills of our parents, we use talk therapy and labeling. Instead, we must work with the real causes of problems, in our environment.

GETTING OUT OF THE HOLE:
What We Need Help With...

Depression is the brain's natural reaction to a dramatically negative experience or a long series of negative events. We often see this in a physical disaster where people are so unprepared for what has happened to them that they seem in "shock," unable to act, with a brain that has no plan of action.

In contrast, First Responders, who have been repeatedly trained in how to react in a serious disaster, are often able to see clearly and act with a plan that is already in place, in their brain, when disaster strikes.

Students often do this automatically. If they are worried about how they may have done on an exam, they may say, "I really blew it on that exam.", even if they think they did well. Because, they have learned from past experiences that sometimes, even if they think they did well, it may turn out worse than they expect. By thinking they "blew it "they are steeling themselves for the possibility that they might do worse than expected. This can be a useful method of reducing our expectations, as long as it is somewhat realistic.

Perhaps not surprisingly, the best students are the ones that worry the most about how they did on exams. In part, the worry is what spurs them to study harder, to avoid the pain of failing, so, they usually make better grades.

But the brain's natural reaction to dramatically negative events is counterproductive. Like a mouse shaken by a cat, we are stunned by reality, we have no ability to act. When those negative events happen repeatedly, they can snowball into a depression. The hole seems too deep to escape on our own.

Most of us, if we get a bad grade or are rebuffed by another, may feel anxious or depressed for a few hours or a few days. But after a few days, the brain itself seems to kick in and bring us out of it. It is vaguely similar to what happens if we get a cold or the flu. We start getting worse for about four days, then the body's own immune system kicks in and we start getting better. If we have had a seriously depressing event, or a long series of depressing events, we may not be able to get out of it on our own. By definition, if we are depressed continually for two weeks, it may be diagnosable as a Major Clinical Depression. Admittedly, that is an arbitrary time limit, but it is probably reasonable. That is when you might need professional help.

Relatives try to use clichés to bring us out of it. Or, lacking any real understanding of what has happened, they may shame us into feeling that it is our fault, that we should pull ourselves out of the hole. That is often society's attitude toward depression. That only makes the hole deeper, along with feeling that no one can help and that no one understands or cares.

This is the time when outside help may be needed—a psychiatrist with a pill or a psychologist with caring and goodwill and CBT. But even professionals are not well trained enough to know how to counter the effects of the disaster in someone else's mind. Too many clients come away feeling that, if professionals do not understand, then no one can help.

By far the most effective means of dealing with such problems is to educate everyone to understand how their mind works and prepare them to understand how their mind reacts *before* the events happen. By giving them examples of how others have had similar problems and *give them a plan, like first responders, to let them know how to react to the problems of life.* But that also needs an understanding—and the education—of the society around them.

The first step, once again, is to train everyone to be "First Responders", to educate everyone to understand the causes of the problems; to give everyone a program in their brain to understand how to react when we encounter a crisis in our life. Without this step, preventing problems will not happen. Learning about the problems others have faced in life is the first step to being a first responder. Knowledge helps protect us from the worst that might happen.

OPRAH: FROM SUICIDAL TO SUCCESS

Oprah Winfrey, a billionaire and one of the most successful, self-confident women in history, recently told of her suicidal feelings in an interview with Piers Morgan. She says that as a young girl, she was sexually molested from age nine to thirteen. She became pregnant at the age of fourteen and was sent away from her family to live with her father.

Oprah's father had no idea that she was pregnant but started out to set down ground rules as soon as she arrived. Her father told Oprah, at the age of fourteen, *"I would rather see a daughter of mine floating down the Cumberland River* (dead) *than to bring disgrace on this family by getting pregnant."* Oprah says she felt she had no choice but to kill herself.

She went through terrible feelings of worthlessness—not because she was a bad person, but because of the emotions associated with the *words* her father used and much of the rest of society used at that time.

It was not even because she was pregnant at 14—it was because society used fear, shame, and guilt as a form of birth control to try to scare teens so badly that they would not want to have sex. Whatever rational reason there may have been for scaring kids away from sex, this fear quickly lost any resemblance to reality and became simply shame.

Oprah's depression did not come from errant genes or a biochemical imbalance; it came from emotions triggered by her father's words. Words associated with emotion control our minds.

Fear, shame, and guilt were the emotions society used to try to control the teen's sexual behavior back then. It was passed on and enforced by parents, media, preachers, and the teens themselves. It has enormous force in people's minds. Today, it may be hard to imagine just how much fear this placed in other people's minds. Oprah herself notes that back then, *"...you were pregnant at fourteen, your life was over, your life was over!"* It was our society's version of the Haitian voodoo curse.

Oprah said she made a suicide attempt by drinking detergent, more as a *"cry for help"* than because she wanted to die. The fact that she had the honesty and courage to tell this story in public will do more good than psychotherapy or anti-depressant pills because it lets others who also feel worthless know that they are not alone, their future is not hopeless, they do not have cancer, they are only human, they can succeed.

One irony of Oprah's story is that she blamed her pregnancy at 14 on her own "bad choices." Really? What kinds of "choices" can a 14-year-old make? Can she choose what language she speaks? Can she choose not to be sexually molested? The sad irony is that society makes us feel whatever happens to us is somehow "our" fault. Even today, Oprah seemed to feel a need to blame herself. Still, we tend to blame the victim, not ourselves as a society.

Society never blames itself. No one thinks, "We had a responsibility to our children to teach them about life's problems, to understand peer pressure, to be able to cope with reality, to make them better parents." Instead, all we gave them were vague warnings—such as watch out! Sex, drugs, and peer pressure. There was no useful information to be found, only fear, shame, and guilt.

Oprah tells of having her heart broken twice in her young life—something almost everyone goes through. When it happens in teen and young adult years, we all feel that no one else has ever been through this pain before; it becomes far more painful than if we had learned to expect that this happens.

Oprah also tells of how a relative betrayed her by selling a story to a tabloid about her, an experience so painful that she went to bed sick with worry over how this might destroy her career. Again, most of us will come out of life feeling like a friend has betrayed us or betraying a friend ourselves. We may tell a friend about something very personal, only to find they told many others.

By talking openly about the problems of her life, Oprah has done more than most counselors or psychologists can do to desensitize others from the pain of life. Psychologists should learn from the value of her stories what is most important as a method of psychotherapy. It all goes back to the work of Albert Bandura. To tell our children about these stories, to understand that we can survive all this and go on to a better life, is the best kind of therapy. Or, better still, preventive psychology—knowledge we owe them in advance. And you do not need a psychologist or a pill to benefit from this simple knowledge; it is desensitizing just to understand that others have survived this.

This is the kind of knowledge that we should be teaching in our schools and what we need to get across to students and adults. It protects them from our cultural version of a voodoo curse. It lets them know they are not alone in feeling so badly; that others have been there and have gone on to succeed and do well.

Therapists could use this kind of information in their own therapy to help their clients. If we fail at this, if we reduce our "help" to nothing more than peddling pills or a trip through psychotherapy, then we have failed. This is what is needed to protect our children and ourselves from the slings and arrows of the value judgments society embeds in our minds.

STATING THE PROBLEM:

No One Seems Quite Able To Put It Into Words

Holden Caulfield, the young man in the late J.D. Salinger's classic, award-winning book, *Catcher In The Rye* is a Shakespearian tragedy in the making. The book is easily the most powerful voice for the average anti-hero teenager in our culture and often required reading in college.

The word that stands out most is Holden's view of people as "fake." Everything he sees around him—love, sex, school, his future, even the teacher he idolizes—Holden comes to see as "fake." At the end of the book, he lies, beaten down by the reality around him, in a mental institution. He even labels himself as "fake." Now society has labeled him as mentally ill.

This is such a powerful book because most teens, at one time or another, feel the very essence of the problems Holden Caulfield felt. Salinger put into words; the angst, fears, feelings of futility, and the lack of a clear future that often plague our youth.

Salinger himself is also famous because, after 1966, he never wrote another work of fiction. He took himself to a mountain retreat where he remained, like a monk in hiding or a modern Henry David Thoreau, until he died in 2010. Why? That may not be as big a mystery as some think. Following the publication of *Catcher,* the book's reception ran the gauntlet between severe criticism for being obscene and his frequent use of dirty words, to being "immature" and censored from the public libraries, to finally adulation that made it required reading in many college courses.

Many speculated, possibly accurately, that the main character was, in fact, Salinger himself. In the end, like Holden Caulfield, Salinger seems to have given up on people, even though people made the book into a sensation. Why did it become a sensation against all odds? I have to think that is because so many readers saw themselves in the troubled youth, so many related to the "fake" they see in society.

> *"There is nothing to writing, all you have to do is sit down at a typewriter and bleed."* Ernest Hemmingway

Literary critics Arthur Heiserman and James Miller said of Salinger's main character, *"It is not Holden who should be examined for a sickness of the mind, but the world in which he has sojourned and found himself an alien. To 'cure' Holden, he must be given the contagious, almost universal disease of phony 'adultism'."* They got it right.

Perhaps the sad irony is that Salinger himself may have died without ever fully realizing that it is society itself that is insane, not Holden Caulfield. Society has failed to give our youth an understanding of the problems they will face. We are the ultimate "fake," immersing our young in what Einstein called, *"glittering words and hypocrisy."*

Albert Einstein, world renown as a genius, wrote in his autobiographical sketch that, at the age of 12, he received the most profound impression that *"...youth is intentionally being deceived by the state through lies..."*

It was, Einstein said, *"...a crushing impression. ...Mistrust of every kind of authority grew out of this experience, a skeptical attitude toward the convictions that were alive in any specific social environment...an opinion that never again left me..."* even at the age of 70, when he wrote those words. Einstein said that this realization let loose, *"...an orgy of freethinking."* Today, like Holden Caulfield, he might end up with a DSM-5 diagnosis of "Paranoid Schizophrenia." And maybe no one would even ask why. Was Einstein the one who was crazy? Hardly.

"PEOPLE DON'T TRUST ANYTHING ANYMORE. WE DON'T TRUST OUR GOVERNMENT. WE DON'T TRUST OUR AUTOMOBILE MAKERS. WE DON'T TRUST OUR PHARMACEUTICAL HOUSES. WE DON'T TRUST OUR DOCTORS OR OUR HOSPITALS OR OUR COPS. WE DON'T TRUST CONGRESS. AND A LOT OF AMERICANS DON'T TRUST THE PRESIDENT OF THE U.S. SO, THE MEDIA ARE SIMPLY REGARDED AS ANOTHER PIECE OF THE AMERICAN ESTABLISHMENT. PEOPLE THINK WE'RE TRYING TO GET AWAY WITH SOMETHING, TRYING TO PULL THE WOOL OVER THEIR EYES, TRYING TO PUSH AN AGENDA." NEWS LEGEND MIKE WALLACE

This comment, by famed news anchor Mike Wallace, was made in 1969. Yet it sounds like today. Are most of us in danger of becoming preoccupied with the "fake" of society, excited only by the thrill of being entertained, unwilling to learn what we need to know to understand what is going on all around us? Are our schools likely to continue to spoon-feed our children pablum instead of reality?

THE PAIN OF THE TEEN YEARS:

Teenagers often complain that they are "unhappy;" that they "do not care" about school or anything. None of this gets at their real problem. They simply do not know how to explain the source of the problems in their minds.

The *real* problem is something they cannot describe, something too vague and too unfamiliar because they have never before heard anyone talk about it, unless they read "*Catcher.*" They feel left out and fear there must be something terribly wrong with them, that they will never be loved or a success, that life holds no hope for them because they can see no clear future for themselves; that they are a failure in a society where they have no hope for success.

They feel no one else could possibly understand. Every time they try to explain, others give them only hollow platitudes about counting their blessings and how lucky they are.

One similar story is from an interview by William Shatner with the rock star success Rick Springfield. Springfield had just written an autobiography opening with his suicide attempt at the age of seventeen. Why would this man, now a millionaire who has achieved so much success in life, as a poet, writer, and singer, attempt suicide at seventeen?

Springfield says he saw himself as a failure at *"the only life arena"* he had ever known—school. He saw himself as unpopular and unliked, without friends, and a failure with girls (all the things on which teens judge their own value). He says he came to feel that something inside of him must be damaged. He had no one he felt he could talk to about his feelings.

Following yet another argument with his mother over his again refusing to go to school, he hung himself in the garage. Fortunately, the rope came loose. He fell to the floor. He saw this as another chance. And his story gives hope to everyone in school who has ever felt so badly.

Although he says nothing about bullying or being put down by others, the refusal to go to school is often the only sign a parent will ever hear of their son's problems. They do not refuse to go to school for no reason or just because it is boring; everybody knows school is boring. No, they do so because the school environment has made them feel so badly about themselves that they cannot face what they see as an ego devaluating experience, a daily failure in their life, the "only life arena they know," that others think less of them, that they are condemned by the indifference of others.

The greatest problem of adolescence comes if, like Rick Springfield, they compare themselves to the other kids they think are more successful. It is the same problem as with Abe Lincoln comparing his relative success as a lawyer, with that of Stephen Douglass, who was far better known, a name renown throughout the country. If we have no other comparison to make, no other criteria to go by, we cannot help but come off worse than others. Incredibly, we don't usually compare ourselves to the average person around us, only to the ones at the top. That is a setup for disappointment.

Teens cannot put their problems into words; they cannot "communicate" with their parents or each other about their fears; like Holden Caulfield, they feel lost, hopeless, and doomed to lose at life and love. They do not see the slightest hope of getting anything out of other people or psychologists, except another handful of clichés and a pat on the back.

Instead, they need counter conditioning; give them stories of all others who have gone before them who felt the same way, yet went on to do well in life despite their early problems. Give them direction in life. Give them an understanding that the value judgments of others are often as fake as Holden Caulfield and Kurt Vonnegut and Albert Einstein realized.

The only way to get these ideas across is by showing them, in stories or movies, how others have been through the same problems and done well, of how the value judgments of others are biased and relative to the culture you grow up in. Tell the true stories of Oprah Winfrey, Abe Lincoln, Winston Churchill, Rick Springfield, J.D. Salinger, and many more.

If we can do this in public schools, if we can get to the young people before their peers or society condition them, if we can teach them to understand the power of peer group pressure and the ideas written into our brains, then we can do much to prevent many of the problems of living.

Virtually everyone in history who has gone on to be a success, who has achieved something worthwhile, has experienced the same feelings of failure, the same fears and frustrations as they have. However, it would never occur to most people that anyone else had ever felt so badly. That is not what they read in our history books. Our schools glorify a very one-sided and positive image, devoid of the reality of life and emotion, without understanding what life is all about.

WE ALL GO THROUGH PAIN & SUFFERING IN LIFE

"The aim of psychoanalysis is to relieve people of their neurotic unhappiness so that they can be normally unhappy." Sigmund Freud

Freud was right about one thing—unhappiness is not a clinical state; it is a normal problem of living. But it can be changed for the better. If we are aware that being unhappy does not mean one is "mentally ill," or doomed to a life of failure, then that knowledge often provides relief, by itself, from the pain of society's labels.

Nearly half of all Americans will "seriously consider" suicide at some point in their lives. Yet, the vast majority of those will never even attempt suicide; they are only expressing a feeling. Pain and suffering are part of life. Education can make it better. Understanding that others have felt so badly and still gone on to succeed is what works.

But that is not what people hear. If a teen or even an adult says he just wants to "kill himself," everyone panics. Teachers, parents, even psychologists no longer become rational. Even professionals constantly warn, "Get them help, immediately!"

Yet, when someone says they do not feel like living, they usually mean they want the bad feelings, the pain of living, to end. When others panic, it sends a message, "Stop talking about it" —you cannot get any help, you only get panic and rushed off for "help." The help consists of pills and talk therapy.

Then, we are often surprised when they may commit suicide without ever telling anyone, "I never knew anything was wrong," or "he was doing so well" we may hear. Of course, they never told anyone—they learned that other people only panic; they do not help, so it does no good to speak out. The hysteria, even by professionals, over "get them help," has created a problem by itself.

One thing that can help is to honestly show them that this is not uncommon; they are not hopeless and doomed to a life of bad feelings. Most people go through similar feelings and go on to survive and do well. If we taught this in our schools, if everyone knew this, it would not be something parents and teachers panic over. We have to give them the knowledge it takes to understand that others have felt so bad and succeeded in life.

From Lincoln to Oprah and from Hemmingway to Edger Allen Poe and Vonnegut, we see a great level of pain and unhappiness in life among the most successful of people. Many look back on the pain of their youth with some degree of amusement over what worried them back then.

Somerset Maugham, the famed author of the dark novel *Rain*, tells with some humor of how a newspaper critic said of his work, "*This man will die drunk in the gutter*" —a reference to Edgar Allen Poe, who, legend had it, did die drunk in the gutter. Maugham's book *Rain* became a classic novel of his day and later a motion picture. Maugham laughed at it, looking back.

In a beautifully honest PBS documentary by Ken Burns about Mark Twain, Burns notes that when Twain was a reporter for a Nevada newspaper, he was challenged to a duel and quickly left for a newspaper in San Francisco. There, he denounced brutality and corruption in the police, leading to a lawsuit. Twain was fired, and his life spiraled into the gutter; he was so ashamed of being fired that he avoided meeting people he might know on the street.

"...I felt meaner and lowlier and more despicable than the worms..." One day, he came close to suicide and put a gun to his head *"...many times I was sorry I did not succeed, but I was not ashamed of having tried."*

Writer Kurt Vonnegut, often called the Mark Twain of the twentieth century, went through enormous pain in his life. His mother committed suicide, and he lived through the firebombing of Dresden as a prisoner of war in Germany, where he was forced to help bury the bodies of those who died in the firebombing. He later told the story of the firebombing of Dresden in his book, "*Slaughterhouse-Five*"

Vonnegut attempted suicide himself after he had already become famous. He recovered. He lived to be 93. The overwhelming majority of people who attempt suicide do not go on to do so again. With a history like that, you might expect he would have problems. He blamed it partly on the stories we are told growing up—the unrealistic heroes of movies, historical figures, the "phoniness" of life, as Holden Caulfield might say. Vonnegut commented on the chaos of life;

> *"...Once I understood what was making America such a dangerous, unhappy nation of people who had nothing to do with real life, I resolved to shun storytelling. I would write about life... Nothing would be left out... It is hard to adapt to chaos, but it can be done. I am living proof of that: it can be done."*

Vonnegut was famous for writing about the pain and suffering of life without being judgmental, without heroes, just the reality of life. His own experiences seem to have given him empathy with the suffering of others that those who have not had such experiences do not have. You see the same empathy and understanding in many; Mark Twain, Mahatma Gandhi, Robin Williams, Anthony Bourdain, Abe Lincoln, Oprah Winfrey, Maya Angelou, and many more.

CONDITIONAL LOVE VS. UNCONDITIONAL LOVE

Why did an incredibly beautiful Jennifer Lopez repeatedly fail in relationships? Twice divorced, failed relationships with Puffy Combs and Ben Affleck, she wrote in her book *True Love* about how she always suffered from low self-esteem, felt unworthy of love, and what changed her. Who would have believed that one of the most beautiful, talented, and successful women in the world would feel that way? Yet that is not uncommon.

She said she learned about unconditional love from her children.

Most of us grow up in a world of "conditional" love. Our parents like us if we are "good little boys and girls." Friends like us if we live up to their expectations of us, if we "act right". Other people like us if we are nice to them. We grow up thinking that "love" is conditional.

Dogs and children love us unconditionally. People, not so much.

World-famous Maya Angelou, writer, poet, sage, had two PBS documentaries about her life. Yet at the age of 14, she says she saw herself as "ugly," "flat-chested," "unlovable." Hearing from others about sex, she wanted to experience what they had talked about. She became pregnant at 14, had feelings of hopelessness not uncommon to teens with far less to worry about, yet went on to great success.

Why? What made her feel so bad about herself? What made her a success? Again, and again, we see the value judgments of others determining our self-worth. In part, she was raised by an aunt who helped give her a sense of purpose. She found that others could identify with her feelings. Her book, "*I Know Why the Caged Bird Sings*" is a lesson in success coming out of failure.

Joseph Biden, has a reputation of sincere concern for the suffering of others. After he was elected President, a newswoman asked him if this concern came from his sad experiences with the death of his first wife and child. Biden said that it actually began from being a child with a severe stutter. Other kids made fun of him for his stutter. That gave him a sincere concern for the pain of others. Pain in life creates a sincere empathy with the suffering of others.

> *"The most beautiful people we have known are those who have known defeat, known suffering, known struggle, known loss, and have found their way out of the depths. These persons have an appreciation, a sensitivity, and an understanding of life that fills them with compassion, gentleness, and a deep loving concern. Beautiful people do not just happen."*
> ELIZABETH KÜBLER-ROSS

HOPE: LEARNING THAT OTHERS HAVE SUFFERED AND SURVIVED AND SUCCEEDED

Winston Churchill, the "English Bulldog" who led England during the desperate times of the Second World War, is the very image of strength and courage in the face of overwhelming odds. Yet, in his autobiography, he spoke of his "black dog" of depression that dogged his steps wherever he went.

In the First World War, he had been the First Lord of the Admiralty in England. He singlehandedly planned and executed the invasion of Gallipoli, which turned into the greatest military disaster in English history up to that time. The English-Australian troops were stopped on the beach with horrendous casualties. Instead of withdrawing the beleaguered troops, he insisted on sending more and more into the disaster. It took months before the troops were finally pulled back. Churchill was forced to resign in disgrace; he felt personally humiliated. Not until the time of the Second World War, did he regain the respect of the people.

"If you are going through Hell, keep going." Winston Churchill

Throughout his life as a lawyer, Abraham Lincoln went through periods when he felt "like a failure." At the age of 40, he sat down and wrote a note, comparing his own failure in life to the greater success of other famous people of his time, such as his competitor, Stephen Douglass, who was renown throughout the country. We see this comparison repeatedly in school when teens compare their own success, in sports, dating, and looks, to that of others. It creates a sense of failure.

Lincoln wrote, "Stephen Douglas's life is one of splendid success, whereas I am a failure, a flat failure."

Michael Burlingame, the author of *The Inner World of Abraham Lincoln,* tells with great passion the story of Lincoln's feelings of failure, which Lincoln himself wrote about at forty. He did not seem to recover from his feeling of failure until he forcefully spoke out against slavery, which made him very popular in the Northern states and very hated in the South.

Only days after Lincoln was elected President of the United States, the Civil War began. In a remarkable book by Joshua Shenk called *Lincoln's Melancholy: How Depression Challenged a President and Fueled His Greatness,* Shenk describes how, during the dark days of the Civil War, Lincoln was often seen prowling the halls of the White House, his head down, hands behind him, saying things like, *"I must have relief from this anxiety or it will kill me."* It is likely that Lincoln himself never thought that history would one day remember him as the greatest American President.

One of the first officers killed in the war was Elmer Ellsworth, a friend of Lincoln. Michael Burlingame describes how Lincoln breaks down weeping in public when he gets the news of his death. This was a scene that would be repeated again and again throughout the war. No human being who cared about others could help but feel a sense of pain and guilt, knowing that he had sent other people to their death in war.

So how did Lincoln deal with such anxiety and depression? He suffered.

There was no alternative back then, people just suffered. There was no one to tell them they are not alone, no examples in their literature about others who went through this and went on to greatness. No pill to ease the pain. Today, we have Lincoln as a great example to let us know that others have made it through. Unfortunately, all of our history textbooks are wiped clean of reality, students learn nothing about life.

Most of us will never have to go through the depths of trauma and guilt that Lincoln and Churchill felt, yet this is the same emotion that real people often encounter in life, even over far less serious events. It matters not one binary digit to our brains whether the issue is one of life or death, or only one of a teen being made fun of by others—the same intensity of emotion is still there. The brain has little ability to put reality into perspective. We need to impart perspective by teaching an honest view of the problems of living instead of heroic tripe.

Reality is censored from our youth. Our schools reduce history into nothing but the gnawed bones of a story that gives us no hint of a reality infinitely more complex than anything we allow our children to understand. Simply telling them the truth about reality would give them an understanding that, when they have to go through similar feelings, they will know that they are not alone; they, too, can go on to succeed.

In Lincoln's day, people did not use the label "depression." They called it Melancholia and considered it a personality type. People did not think less of Lincoln because he broke down crying in public when he received the news of how many soldiers had been killed at Shiloh; it made him more human. Their own losses in the war made this easy for them to understand.

It is different today. There is a heavy bias against those who psychology labels "mentally ill." In the 1950s, we changed the name from Melancholia to Depression, and it went from being a personality type to being a "disease." We made psychological problems into a "disease" because we wanted to be more like the medical profession. No one thought about the harm it would do.

You can see the harmful effect this has had on our perception of reality by what happens to a politician today who is tagged with this label. In 1972, Senator Thomas Eagleton of Missouri was picked to be the Democratic nominee Senator McGovern's running mate. If elected, he would be America's Vice President. Very soon, the political hate machine discovered that he had been hospitalized for depression.

At first, the Democratic nominee refused to consider changing his running mate. But, after a month of hearing attacks that said, in effect, "How could this man, who has suffered from depression, be allowed to be a heartbeat away from the Presidency?", Eagleton was forced to step down and was replaced with another candidate.

Could Winston Churchill or Abe Lincoln be elected as a world leader in today's society? Or would they be so damaged, so savaged by the political hate machine and a sensation-seeking media, that they could not be elected dogcatcher? Judging by the quality of such people in the past, we should probably rewrite the Constitution to require that whomever we elect to high office should have a history of depression. However, we can see in what happened to Senator Eagleton just how badly these labels have harmed people.

You can see this again today, in that soldiers from the Gulf War who have symptoms of PTSD or depression do not go in for treatment; they do not want to be labeled as diseased or seen as "weak" by their fellow soldiers. The same type of conditioned fear results in male children drowning for fear they might be considered "chicken." We do more harm with our labels than we do good. The very concept of "mental illness" is a damning label in a society that attacks people for being "different" or "weak;" whether we intend this to happen or not, this is the result of our labels.

Andrew Solomon, the author of *Noonday Demon*, said, *"It would be wonderful if we could spare people the experience of depression, but I do think that it catapults people into profundity and moral purpose and I think it does make people gentler and softer and it intensifies all of their emotional experiences and it gives you, at its very best, a great sense of purpose."*

THE POWER OF COUNTER CONDITIONING:
What led a psychologist to the depth of depression to the success of a lifetime and back to depression.

One of the all-time greats in psychology was O. H. Mower. As a graduate assistant of Max Meyer at the University of Missouri, he was enrolled in a sociology course entitled "The Family," taught by a popular professor, H. O. DeGraff. Mowrer's curiosity led him into a forbidden topic of the era. He constructed a questionnaire concerned with trial marriage, divorce, and several questions on premarital sex, types of sexual encounters, and more, as information for his thesis. Meyer and DeGraff looked over Mowrer's paper and approved the study. This questionnaire was to be distributed to 600 students.

The local press heard of the type of questions being asked and raised a cry of outrage against this "immoral" questionnaire, which they believed "undermines the American family system." The college's executive committee fired Meyer and DeGraff, along with whatever passed for academic freedom back then. Meyer's sentence was reduced to one year's suspension without pay, but DeGraff was not reinstated. Mowrer survived relatively unscathed, being only a student, but the experience seems to have left some mark. Of all that Mowrer wrote, he never again wrote about sex. Like Mark Twain, he censored himself.

Mowrer wrote many successful books on *Learning Theory and Behavior* and *Learning Theory and the Symbolic Processes*. Mowrer went on to become interested in the psychoanalytic ideas that were popular then, including Freud's concept of catharsis, what Freud himself called "the talking cure" —a term first used by Freud's client, Anna O. Years later, after he had become famous and successful, Mowrer was worried by self-doubt, feelings that he was a failure, and most important of all, by things he had done in his life that filled him with feelings of guilt and shame, things he had told no one.

Mowrer's colleagues voted him to be the president of the American Psychological Association. Before Mowrer took the position, old feelings of guilt from his youth began again over his early experiences. He could not get the guilt feelings out of his mind. He felt he had to tell someone about his guilty secret. In a moment of candor, he told his wife the truth about the fearful secret of what had happened in his youth. To his surprise, she was very understanding. She supported him. She did not react with shock or horror; she was accepting.

Mowrer felt instant relief, like the weight of an oppressive force had been lifted off his shoulders. Like a cancer victim pronounced cured. The dreaded secret was no longer something to dread. He was free. He was cured—catharsis worked. Freud's "talking cure" worked.

Mowrer felt such relief after the event that he made it a secret no longer. He went about telling others of his guilty secret. But something happened—others did not react with understanding. Their mouths dropped. Their expressions changed. Their reaction was nothing like the acceptance he had received from his wife. Their reaction was one of shock and disdain. You could see it in their expressions. Cancer returned; he was no longer cured. The fears came back with a vengeance.

Mowrer went to the American Psychological Association and refused the presidency. They patiently declined to accept his refusal. However, within weeks, he was hospitalized with severe depression, unable to work, unable even to carry on his life. He remained in a mental hospital for a month.

What happened?

Guilt, said Mowrer, is the cause of psychopathology, he said in his book, "*The Crisis in Psychiatry and Religion*". It is not the only thing, of course, but something important is illustrated here. First, the guilt was not caused by anything he had done, but by other people's reactions. Second, and most important, *catharsis* never works in anything like the way Freud described. Telling your guilty secrets does not free you or relieve psychological tension and does not reduce hydraulic pressure in the mind. No way.

Still, there is something more important. What gave Mowrer such relief the first time was the fact that his wife was accepting. She did not react with horror or disdain. This desensitized Mowrer to believe that his secret was not so bad; that he was a good person after all. When psychotherapy works, it has the same effect.

Talking about his problem is not what relieved the fear, not at all. He experienced relief only because his wife reacted with positive emotions. This works. This is successful. It allowed counter conditioning and desensitization to occur, just as it did when Bandura showed that children's fear of dogs could be reduced by letting them see happy children happily playing with happy puppies. This is something to learn from.

The failure came when he mistakenly assumed that telling others would also relieve his fears. It did not. Why? Because others reacted with shock and disdain.

Others produced fear and self-loathing all over again. Other people's emotional reactions told him that he had cancer. He was a leper. It was a conditioned (learned) emotional reaction. The fear of "what other people will think" is extraordinarily powerful.

It was not true, of course. Mowrer did not have cancer, and he was not a bad person; he was a good, successful person who made a major contribution to human understanding. The point should not be lost that it was not the "reality" of Mowrer's guilty secret that caused his pain; it was how other people reacted to him. One more case of our society's voodoo curse—the fear of what other people will think.

Moreover, what mattered was how he came to think of himself because of how other people had reacted.

Other people are not the criterion we should use in judging our lives. Their ignorance must not become our pain.

What was O. H. Mowrer's guilty secret? He learned his lesson. He never again told anyone else. He had learned his lesson the hard way, from other people's subtle reactions.

GUILT BY ASSOCIATION: The Great Curse of Society

Mowrer's "guilty secret" seemed to be having a homosexual experience as a boy. Back then, homosexual behavior was considered vastly more "sinful" than today. Today, we are only now hearing stories of how boys in the Boy Scouts of America were abused, and the organization itself covered this up for decades. Now, the Boy Scouts of America have declared bankruptcy, unable to pay off all the lawsuits against it.

And we hear in the media of the priests who molested boys. In 2018 alone, just in the state of Pennsylvania, the District Attorney announced that some 300 priests had sexually molested over 1,000 children over decades, and the boys who did speak out were often shamed for doing so.

Just after the Pennsylvania announcement, Germany announced their findings that in their country, over 3,600 minors had been sexually abused by over 1,600 clerics over 70 years. Most were too ashamed to speak out or were not believed when they did. So, they suffered for decades from the shame, guilt, and censorship our society had imposed on them.

Just after these findings, President Trump announced that he did not believe a woman who accused his nominee to the Supreme Court, Brett Kavanaugh, of sexual abuse because "...*if it had happened, she would have spoken out decades ago.*" Yet, she said that she was afraid even to tell her mother that it happened. She was afraid because of her own guilt over what her mother would think if she knew she'd been to a party with boys and alcohol at the age of 15. Despite a lifelong feeling of guilt, she went on to have a successful career as a psychologist.

In sexual assault, the failure of those who have not respected other people's rights is not their failure; it is a failure of our society to educate the young to value others' rights, and more important still, our failure to interfere with the harm done by the peer group that spawned the problem. Instead, we allow the peer group, the male subculture, to value "getting sex" as more important than other people's feelings.

Ten years earlier, when similar problems happened, no one talked about it publicly; church elders had covered it up in shame. These children were ignored or afraid to speak out back then, yet several of those who did speak out, thirty years after the incidents, committed suicide. Why?

One would think that the discovery of just how many children had been suffering from "guilt and shame" over this long would have been somewhat desensitizing, given that they had long felt alone, as if they were the only ones. Yet the reaction of the press made the desensitization less valid.

They did not commit suicide because of something that happened thirty years ago; they committed suicide upon hearing their experience being described with "shock and horror," as one newswoman called it, from the way we react to hearing about this today. "Despicable" is a word used by a famous woman television personality who spoke with great anger about this behavior on her news show.

When the media and others express such shock and horror about molestation, incest, or other issues, they have no intention of doing further harm to those who were innocent. Yet that is precisely what happens. It is guilt by association. Oprah Winfrey was not suicidal at 14 because of the sexual molestation or even being pregnant; she was suicidal because of society's emotional judgment, reflected in her father's words.

Like the girls at Salem, pointing at the accused, screaming "witch, witch" the adrenalin shoots into the blood, our heart jumps, we are condemned by the emotions of others.

In the more recent trial of Penn State football coach Jerry Sandusky, the adults who told of being molested by him when they were children often broke down crying on the witness stand as they described what happened to them. They were not upset because of what happened decades ago; they were upset due to the continuing guilt and shame created by the hysteria of people talking about how shameful and despicable Sandusky's behavior was. It was, again, guilt by association.

If people had simply responded rationally, saying, "this should not be allowed to happen" instead of reacting with hysterical pronouncements of emotionally charged words like "despicable" and "horrific," then the incident itself would not have been so traumatic. We make it worse by our value judgments. Even when the press does a professional job of reporting the story, society itself continues to be judgmental.

The guilt is further exacerbated by the fact that most of the cases are seduction, not forcible rape, although the legal system makes no distinction. Yet, there is a psychological one. It is common to find that the child often does enjoy the experience at the time. That is a natural reaction. Now, that feeling is associated with something that society is saying is very bad—even despicable. These emotion-triggering words are associated with the child, having been a part of it. As one individual who was molested as a child put it, "*It* (other people's reaction) *made me feel as if I were trash, the lowest form of scum, one of 'them."* That is "guilt by association."

Instead of reacting with reason, we react with emotion. This anger has taken over the media as "justifiable anger" expressed almost routinely when such issues come up on cable news shows. Emotion, not reason, has become the *soup du jour* for the media. For many, we have arguably done more harm with our emotional outrage than was done by the people who molested them in the first place.

We have labeled them, like the label of "cancer." It matters not at all if it is true or not, nor does it matter that we never intended to make them feel that way; we have made them feel that way because of society's emotional reaction. It is simple guilt by association, emotional conditioning. It is, again, our society's version of the Voodoo curse.

Rape is a word that elicits the same emotion as "cancer." If it is applied to you, the emotion overwhelms any rational response to the issue. It becomes the same as a voodoo curse, and society makes it so by our overreaction to the word.

It does not matter whether or not you have cancer; if a doctor tells you that you have it, you will react accordingly. Similarly, it does not matter that these individuals have nothing to feel guilty for, or even if those who react with emotional outrage have no intention whatsoever of making you feel bad; if society labels you with words associated with an emotion, your mind will react accordingly. We have told people they have been associated with something "despicable," and we have done immense harm with that simple association.

Women's groups even object to using the word "victim" —it is just one more label.

> *"Don't paint me as a victim; I am much more interesting than that."* Glenn Close, The Wife.

Society is fairly good today at demanding that molesters be held to account for the harm they do to innocent people, but there is never anyone to hold society responsible for the great harm society does to the innocent by our failure to deal with these issues rationally, instead of emotionally, and by the failure of our educational system to teach our children why they need to respect the rights of others.

Society should be the ones who feel guilt and shame over our failure, but society never blames itself for anything; it is only good at blaming others.

If you want to understand why people behave as they do, if you care to find out why issues raised by the #MeToo movement are so incredibly common, if you want to know why the thousands of children are molested by priests or abused by Bill Crosby and Hollywood moguls such as Harvey Weinstein, or that 26,000 cases of sexual abuse were reported in the military, you have to look at society and our failures:

- The peer group that encourages the abuse of others, bullying, sexually predatory behavior, ignoring other's feelings, all happens because society has failed to control what happens in our school system when we allow emotional conditioning of the peer group, by the peer group.

- A failure of society to use the existing educational system to educate the young to understand the harm they can do with words and the good they could do with understanding and compassion.
- Society itself has encouraged the use of emotion instead of reason, even in our politics.
- And fourth, we cannot allow the peer group in school to condition each other not to be a "tattletale" to fear the peer group's opinion more than they value doing what is right. In our schools and in our politics, going along with our peer group overwhelms our obligation to do what is right.

SOCIETY MAKES MEN AND WOMEN DIFFERENT: SEX AND SHAME

If you want to understand how powerful society is in making men and women different, look at how males and females react to sex. We have just been through a #MeToo reaction to stories about politicians and others allegedly attempting to force sex on a 15-year-old girl. The very idea of an older man having his way with a teenage girl strikes us as gross. We occasionally hear of a female teacher who has sex with a male student. The media only covers that because it is sensational.

Yet, if you reverse the situation, where an older woman or a female teacher has sex with a teenage boy, most teenage boys would be like, "High Five! Way to go! Wow!" from the other boys. How could something so traumatic to females be so different if it happens to a male? Again, the emotions we hear associated with an idea determines our reality; males grow up in a different reality than females.

For males, having had sex gives them some prestige in the male subculture. For females, society associates sex with shame and guilt. Their reaction has been just the opposite of males. Some say that is changing. Not yet.

Words, associated with emotions, control our brain's biology. Reality is lost in translation.

So often we hear of someone who has had a "nervous breakdown," committed suicide, or been hospitalized for treatment, yet there is no hint of "why." More often than not, it is presented as if it just suddenly happened, for no reason. It doesn't just suddenly happen for "no reason." More likely, people learn not to tell the truth, not to give reasons. If they do tell, people are likely to shame them for the reason, that it is not important enough, or, if the reason is like that of Mowrer, react with shock or disdain.

So, reasons for nervous breakdowns go unstated. Reasons for suicide go unmentioned in the press. They have learned, like Mowrer, not to talk. You see this today in the #MeToo movement, and in the millions of women and men who have been through this; society has made them ashamed of their feelings.

Yet, the reasons are there. Emotional conditioning is our culture's voodoo curse. People may be able to tell their guilty secrets to a therapist who reacts with compassion and understanding instead of horror or derision. This may help to desensitize them to the fear that others will think less of them. That is not the feeling they get from society. Instead of using a therapist, we need to use the entire existing educational system to create understanding, to counter the effect of the emotional conditioning of the peer group by the peer group; and by society.

CATHARSIS OR COUNTER CONDITIONING?

Albert Bandura said that when psychotherapy is successful, it is because it allows some degree of "desensitization" to occur. I have no doubt he was right. The experiences of O. H. Mowrer showed this in dramatic form. Different forms of psychotherapy attack the issue with different methods, but all must deal with the emotions society has embedded in our brains.

An army counselor recently said in a newscast talking about Post Traumatic Stress Disorder that his purpose was to act as a cathartic agent, to allow the release of emotion. In the form of telling others of our problems, catharsis does not release our built-up energy and does not often work in the real world. Many have bought into this idea, accepted by some hopeful initiate who went on to tell the secrets of their soul, only for other people's reactions to beat them down.

One of the ideas often heard from the human potential movement of the 1970s was that people are uptight, conflict-ridden, and fearful because of their inability to disclose themselves to others fully. The theory was widely put forward that if you simply were honest, open, and freely told your feelings to others, you could become freer. "Let it all hang out," "spill your guts," became the mantra. Honest feelings, emotional release, and confrontation with one's problems have all been touted as magical releases from the mind's problems. By ridding ourselves of these hang-ups, we were told, we could release the energy and solve our minds' problems. This idea has been popularized in many forms of therapy.

One young lady, a product of an "encounter group," was so convinced that she should hold nothing back that she went about telling all who would listen, of her sexual molestation, depression, and feelings of guilt. She was honest and sincere and a true believer in the power of therapy, but she was destroyed by other people's reactions. No one wanted to hear it; no one even wanted to be around her. Other people's reactions intensified her depression. She was made worse by her belief.

What may work on an analyst's couch or in an encounter group of understanding people does not work in the real world; ordinary people are not so understanding. Our educational system has failed us. Our media has failed us. They are part of the problem.

Yet to this day, many psychologists and counselors who know much about therapy but little about people, believe that talking to their clients is all you need to do. A more effective method would be to assign homework, readings, or even self-help books to their clients to advance the therapy. Use the stories of Oprah, Lincoln, Vonnegut and more. Use the literature of Salinger and Vonnegut and more, as instruments of desensitization.

If we use examples just described or videos showing how others have been through such problems, survived, and succeeded, we could be more successful; not just as psychologists, but as teachers and parents.

THE GUILT OF THE INNOCENT

One of the most unfortunate psychological problems is one observed by psychiatrists following World War II. During the war, police arrested people in Germany for little or no reason. Some people were turned in by their neighbors for not being sufficiently patriotic, or they were charged with treason for criticizing the government. College professors and students who were not sufficiently pro-German were jailed for opposing the state. Most were there because of their race or religion. People spent years in concentration camps. Nine million people died in Nazi concentration camps. Many survived.

Psychiatrists found a new phenomenon among the survivors—they felt guilt. No one else seemed to.

The Nazi leaders in their public and private statements repeatedly excused their behavior as "necessary" to save the nation from criminals. They said, "We were only following orders." And "We were only enforcing the laws".

The soldiers who carried out the executions and brutality at the concentration camps rarely reported guilt; they excused their actions, saying, "We were only doing our job." "We were just following orders." "We had to do what we did." Everyone had an excuse. Everyone could rationalize their behavior.

It was the innocent victims who felt guilty. Why? How could such a thing come to be?

Psychiatrists came to call this "survivor's guilt." It is not just common to those who were innocent victims in concentration camps; it is also found among rape victims, those involved in an incestuous relationship, and many soldiers returning from combat with Post Traumatic Stress Disorder. Through no fault of their own, they often report feeling intense guilt.

Women's groups often note that police questioning in rape cases tends to intensify guilt, making the victim feel they are at fault by asking questions about the way they dressed or acted. Most rapes go unreported. The psychological toll on those who do report rapes is sometimes high.

The fact that today, many decades after it happened, we are only now beginning to hear stories of sexual abuse by Bill Cosby, Harvey Weinstein and many others is testimony to just how ashamed and helpless people felt when alone, when the media allowed no one to talk about it in public.

In Pennsylvania alone, over 1,000 children were abused by some 300 priests who were protected by their superiors. The victims were often made to feel ashamed that they would complain because no one seemed to believe them.

So why do innocent people feel such guilt? In part, because our society has inflicted its ignorance and superstitions upon victims. All our lives, we have heard that we are responsible for what happens to us. We hear that we must "pull ourselves up by our own bootstraps," that we must "take responsibility for our own behavior," and if something happens to a child, adults often question them until they find some blame to place on them for what happened. It is as if we feel that the child must be cautioned not to let it happen again by blaming the child's own behavior.

Parents may feel the child must be shamed to "teach" them not to do something again. They may think teaching them to take responsibility for their behavior is what it is all about. But this has a negative effect far beyond any "teaching" this might provide; it generates unforeseen consequences. Shaming children may lead them to blame themselves, to think it is their fault, as Oprah blamed herself for "bad choices." You see this in many people in the #MeToo movement, who felt so ashamed of something that was not their fault that they, like Mowrer, were afraid to even talk about what happened to them for decades.

Many of us hear from childhood that bad behavior is punished; that God rewards the good and punishes the wicked. That Santa does not bring gifts to "bad" little girls and boys. So, if something terrible happens to us, we may come to believe that we have no one to blame but ourselves; something we often hear from our parents.

Guilt is not what we have been told it is. Even a dog will lower his head and look guilty if he is yelled at. The guilt is only a fear of how others will react. Yet, it is as if the dog feels guilt because what he has done has caused such a bad reaction from others. People react the same. Not knowing any better, the individual thinks it must be something he or she has done that makes them feel so bad.

Even if people do not blame themselves, even if they know better than to think it is their fault, there is always the fear that other people will react that way; that other people will think less of them because of what happened to them.

It is nonsense, of course; bad things often happen to innocent people. But our societies implant superstitions in our minds. The idea of self-blame is a part of our minds. And sometimes, we should blame ourselves. Yet, no matter how irrational, we often believe we are to blame for things we should not feel guilt over. It takes a good degree of knowledge to understand just how superficial our value judgment is. It takes desensitization to change the fears that plague our minds.

No one should ever have to feel shame or guilt over what others have done to them, especially when it is done by society itself. Never in human history has there been a more bizarre example of the problems created by society's ignorance in using conditioned fear to control the minds of the young.

"Insanity in individuals is something rare – but in groups, parties, nations and epochs, it is the rule." **Friedrich Nietzsche**

What happened to the survivor guilt in the concentration camps, or to O. H. Mowrer and Oprah, and thousands of children abused by priests or women who were sexually abused or molested has never been a more egregious example of the ignorance of society. "Never again" should be the mantra. Yet we teach nothing about this to our children. We do nothing to try to help.

"If there is life on other planets, then Earth is the Universe's insane asylum."
VOLTAIRE

SO, WHAT WORKS?
Desensitization and Education

So, what does work? Desensitization works. Knowing the difference between society's superstitions and reality works; arming yourself with knowledge works, teaching the entirety of society to be understanding works. It sounds too trite to be true, but bear with me for a few more chapters, and I think you will understand, from your own feelings and emotions, just how effective it can be at controlling the forces of mind that everyone experiences to one degree or another.

In one of the studies mentioned earlier, Albert Bandura, using the same techniques as Pavlov and Mary Cover Jones, went on to discover that children who were already terrified of dogs—could counter-condition their fear by showing them *videos* of a *happy* boy *happily* playing with a *happy* puppy. The emotion of *happy* countered the fear of the puppies, allowing them to pet a puppy for the first time. This would turn out to be perhaps the most important discovery and one of the least known in psychotherapy.

By the same method, *seeing videos or reading stories* of other people's problems can help desensitize our fears when we see that others have the same problems and have gone on to succeed. This is why the stories of Lincoln, Oprah, and so many more are important to give to ourselves and our children.

It is not sufficient to wait until these problems develop and then try to cure them. We have failed as a society, as psychologists and professionals, as teachers and parents, if we do not take the necessary steps to teach children and adults to understand these kinds of problems *before* they occur; to prevent the problems in the first place. These are the kinds of problems that can easily be prevented. And they could be prevented with only a very tiny cost and only a modest change in our educational system.

Children should not have to learn how their mind reacts the hard way, as we have. The experiences they have may set the course for the rest of their lives. Instead, they can learn by being told of the ignorance and foolish mistakes that we, the adults, have made in our past. And by such learning, they will be less likely to be caught by the same forces. Never assume that telling them once is enough; it will take many different stories from many different sources to overcome the force of our peers.

To understand what makes for success in life, we must also understand what causes our problems. What are the mysterious forces that cause the psychological problems of living?

"LET PEOPLE CLEARLY REALIZE..."

The number one cause of psychological problems in your life or mine is not genes in our DNA or biochemicals in our brain. The overwhelming cause of most psychological problems is *other people*. Think about what the major causes of problems in your life or the lives of others are. Problems come from our interaction with other people—our parents, bosses, friends and society. Other people make life positive or negative for us. And sometimes we are the "other people" who make life good or miserable for others.

Perhaps the most important comment ever made on the causes and cures of human problems was from psychologist Abe Maslow, most famous for his Hierarchy of Needs;

"Let people clearly realize that every time we hurt or humiliate or belittle another human being, we become forces for the creation of psychopathology, even if those be small forces. Let people clearly realize that every time we are kind, decent, psychologically democratic, we become psychotherapeutic forces, even if those be small forces."

ABE MASLOW

That is the single most profound comment on the causes and cures of human problems I have ever heard. Maslow may have made a small mistake by using the term "psychopathology," as that would include even biologically based disorders such as schizophrenia or autism, but no one has ever gone to the meat of human problems as Maslow had.

This is psychology's version of "Do unto others..." and "Judge not, that ye not be judged." Along with a reason. Yet, to be effective it must come with example after example to be able to get the ideas across. Otherwise, the importance will be lost in a wayward cliche.

Yet, the press ignores this, not sensational enough to compete with the magical-mystical words like "DNA" or "magic pills" and ignored by our educational system who should be teaching this to everyone.

CHAPTER 4

THE FIRE IN YOUR BELLY - SUCCESS

The Genesis of Human Success

What makes for success in life? The one outstanding comment I see throughout many successful people's lives that we discuss in this book is summarized in the words of author and Supreme Court Justice, Oliver Wendell Holmes, who said:

"Life is a romantic business, it is painting a picture, not doing a sum, and it will come to the question of the fire you have in your belly."

What he meant by "the fire in your belly" was the emotion that directs our lives, the enthusiasm that leads us to acquire knowledge and to work toward a goal, and the strength to overcome life's problems that others dump in our path. As we read the stories of great successes, we find this theme repeated again and again.

Albert Einstein spoke of the *"sense of wonder"* he experienced as a child when he saw some unseen force move a compass needle. This electro-magnetic effect became the subject of his first papers.

Nobel Prize-winning mathematician/philosopher Bertrand Russell spoke of his first thrilling childhood experience with mathematics, Euclidian geometry, which his older brother taught him, as being *"...as delicious as first love."* Math? As delicious as first love? Huh? He went on to win the Nobel Prize in mathematics.

Astronomer Carl Sagan, from his famous series *"Cosmos,"* spoke of how, as a child, he first learned the astonishing fact that all the stars in the sky are really suns, like our sun, and that these, too, might have planets and life. This feeling of awe was part of what led him to a lifelong interest in the subject that made him a success.

Einstein spoke of, *"...an orgy of freethinking."*

Nobel Laureate Jean-Paul Sartre, awarded the Nobel Prize for literature, tells in his autobiography, *The Words,* how on seeing that young Sartre was drifting without direction, his grandfather took him aside and told him he had the makings of a great writer, like the writers of the books on his grandfather's bookshelves. This gave Sartre a new direction in life, leading him to see becoming a great writer as his new superhero.

"IT FLOWED INTO MY BONES; IT DIRECTED MY LIFE..." **THE WORDS,** JEAN-PAUL SARTRE

"IF YOU DON'T KNOW WHERE YOU ARE GOING, ANY ROAD WILL GET YOU THERE." LEWIS CARROL

EMOTION RULES GENIUS

Historian-philosopher Santayana said, *"We only become good as adults at those things at which we have played at as children."*

Many of the great minds in any subject, art, music, or sports speak of that spark; that intangible enthusiasm of youth that had a profound effect on the rest of their lives. Yet none of them explain the origin of that spark. We will examine the events in their lives that led to their success later in this book.

Our culture trumpets the need for better schools and more education, extending the school year to eleven months instead of nine, and holding students and teachers accountable for success in math or reading scores. Many parents are intently aware that academic skills are important for theirs and their child's future; they buy them educational toys, tutor their own kids, or even hire tutors to enhance their child's education, to give them a better chance at grabbing life's brass ring. Adults may go back to college to gain new skills. But the single spark missing from discussions about what we need to do to save our public educational system is nowhere to be found in our leaders.

What Justice Holmes called the *"fire in your belly,"* Albert Einstein called a *"sense of wonder."* No one described it better than Nobel Prize-winning physicist Richard Feynman, who spoke of the *"tremendous excitement"* he felt at knowing, even for a short time, something about the way nature behaves... that no one else in the world knows.

It is an emotional excitement over *knowledge for the sake of knowledge* that seems identical in every respect to the emotional excitement of youth's games and sports. Goethe said, knowledge and ideas had always been a game to him—one he forever delighted in playing. But don't make your child into someone who would not have the most important people skills.

What is the origin of the *"fire in your belly"*?

THE SOURCE OF THE FIRE IN YOUR BELLY

Years ago, I wrote of the origin of that spark in a story that seems to have little relevance to success at first. It was about a man who loved thunderstorms as a kid and felt a sense of exhilaration around them, even as an adult.

The early years of childhood are always important. But it is not that these early years are more important than later ones; it is that the first few experiences we have with anything tend to set the passion for much that will follow.

Thunderstorms are intensely frightening, and many children are terrified of them. Dogs often try to hide under the bed to avoid the "boom" of thunder. Instead of being afraid of the roar of thunder, this man would feel an emotional thrill at the approach of a thunderstorm. He would seek them out; he loved to smell the air, listen to the rumble of thunder, and hear the rain coming down.

Why? How could this be? What could take something as intensely frightening as a thunderstorm and turn it into something so exciting? Was it embedded in his DNA? Was it a product of his unique biochemical imbalance? Was he just born that way?

He had no awareness of why he felt this sense of exhilaration over thunderstorms. He would never have known how this came to be, except that his mother told him why at the age of 25.

His mother had been terrified of thunderstorms as a child, and she was determined that her son would not grow up with the same fear. So, when he was about six years old, she took him out on the porch when a thunderstorm approached way in the distance. "Oooh, look at the pretty lightning! Listen to the rumble of the thunder."

"Smell the air!" She said with enthusiasm. "What is that in the clouds? Is that a dog?"

Using nothing but the positive emotion in her voice, she was able to take something as profoundly frightening as a thunderstorm and turn it into a positive emotional experience—an experience that lasted a lifetime.

First experiences set the emotional tone that determines how we react in future experiences. Saying that is not nearly so important as giving an example. The following is from a story by a college professor that I wrote about in *PSYCHOLOGY: The Science of Human Behavior*, many years earlier.

All of my life, I have had a thing for thunderstorms. Some people are terrified of them, but for as long as I can remember, I have found them exciting and exhilarating. I could sit on the porch for hours listening to the rumble of thunder in the distance, breathing in the rain freshened air, and watching the lightning dance across the sky ... I can remember having been surprised to find that not everyone shared my fascination with thunderstorms; it seemed so natural, but it was not until late in life that I learned the origin of my interest.

My mother was herself frightened of storms. But when she became a mother, she decided that her son would not grow up with the same fear. So, when I was very young, every time a storm was brewing in the distance, she would take me out on the back porch and introduce me to the thing that frightened her as if it were the most fascinating discovery that we had made:

> *Mom: (with feeling) Ooh! Listen to the thunder in the distance!*
> *Me: (with empathy) Yeah! Thunder!*
> *Mom: Mmmmmm! Smell the air!*
> *Mr: Yeah! Air!*
> *Mom: Look! Look at the pretty lightning!*
> *Me: Wow!*

Actually, I remember absolutely nothing of the events. If my mother had not told me about these little sessions years later, I would have no idea why I have such a feeling of exhilaration associated with thunderstorms. Yet to this day, thunderstorms give me a vague but heady sense of power.

Years after this was published, former First Lady Betty Ford wrote of an identical experience in her autobiography. Betty Ford's mother was also afraid of thunderstorms and did much the same thing for her, with the result that, even at the age of 70, Betty Ford says she felt a sense of "exhilaration" at the approach of a thunderstorm.

PSYCHOLOGY'S ATOMIC THEORY:
The Fire in your Belly

That is power; think about this. A parent can take something as frightening as a thunderstorm and turn it into something exhilarating; a feeling that can last a lifetime.

That is a tool that we call a Conditioned Emotional Response in psychology. It is a tool that few parents are aware of; even few psychologists are aware of its role in determining our life course.

Motivation is to psychology what nuclear power is to physics. The first nations whose teachers and parents realize its importance will have a major advantage over those where people cling to the glorification of entertainment.

So, how does this apply to success in life? We all spend time and energy doing those things we have come to see as exciting. It could be any stimulus; public speaking, the mystery of the movement of a compass needle, the delicious texture of math, the awe of the universe, or sports. Any Stimulus can come to elicit Any Response (almost) already wired into our brain—especially emotional responses, the very basis of our motivation.

Pleasure, fear, and anger are the three dominant emotions we are born with. Everything else—shame, guilt, anxiety, pride, jealousy—seem to be a derivative of these emotions, a conditioned emotional response that varies in degree or overlap.

These emotions are associated with stimuli—thoughts, words, and images in our brains. The stimuli we see or hear from others trigger our emotions. Words and thoughts associated with them come to control the emotions in our brains.

Our culture already uses this technique. We take something profoundly dangerous that causes more injuries every year than almost anything in our society, and we glorify it with applause, cover it in accolades, cheer it in person, and our networks are a slave to its glory—sports.

Sports are painful. Over 250,000 young people end up in emergency rooms in hospitals each year due to sports injuries. High school and professional football players often have brain concussions, paralysis, and lessor injuries that may haunt them for the rest of their lives. Yet we glorify this with clichés such as, "No pain, no gain!" A 2010 study of former NFL linemen, the 300-400-pound blockers and tackles in the football league, found that half of them have heart disease by the age of fifty.

The length of the average career in the National Football League (NFL) is only about three-and-a-half years. Most are quickly replaced due to injury, or they are replaced by some young buck who is quicker or hungrier than they are. The joke among NFL players today is that NFL stands for Not For Long.

How is it that we have taken something so painful and dangerous and elevated it to the status of a goal for kids that dominate their thoughts, lives, and passions? How have we made this into something they seek to achieve in the fires of youth and enjoy watching, even into their old age?

What begins at home with the enthusiasm of dad playing catch with junior and the cheers of the little league is institutionalized into the school system. What is it that high school glorifies? Not knowledge or education—only sports. We have weekly pep rallies in school, totally dedicated to glorifying football. Crowds madly cheer when "their" team gets the ball and scores. Cheerleaders hop up and down with enthusiasm and short skirts, all devoted to sports. A drill team is devoted to sports.

Majorettes prance around the stadium at half time. The band plays...

Men who played catch with their dads and who listened to the wild cheers of their parents in the bleachers as they played in little league will experience a sense of exhilaration over baseball for the rest of their lives, even past the age of sixty.

For their wives, who rarely played catch with dad or played in little league, watching baseball is like watching their toenails grow; "it's the bottom of the 9th, and the score is tied 2—2."

How many high school students would ever brag to other students that they just won the *National Computer FreeCell Contest*? Few would dare. That would be a hoot!

Most students consider playing solitaire on a computer as trivia; yet, is it any more trivial than winning the *National Catching A Football Contest*?

What our society glorifies is what students yearn for. They want the honor of being on the "team." We go out of our way to glorify football, baseball, basketball, and even golf. They want the honor of getting a letter jacket just for showing up. Girls like the guys with the sports letter jackets, not the guys with the Computer FreeCell Team T-shirt.

We make sports exciting, something we do with our dads and our friends. We make education and homework a painful duty you must do alone. Small wonder most students see no thrill in knowledge.

Which American is the world's greatest swimmer? Most can quickly guess it is Michael Phelps, a five-time Olympic gold medal winner. Okay. Now, who is the number two greatest swimmer? Who is number three?

If you are number one, you are everything. Michael Phelps is the one who got the front-page pictures, the adulation of the masses, and 40 million dollars in product endorsements.

Who remembers number two?

If you are number two, you are not squat.

Is this the reality we want our children to learn?

THE KEY TO SUCCESS IN OUR SCHOOLS

Every President has insisted on the importance of education. Stay in school, is the mantra. The key is not what the President of the United States tells students is important; it is what students tell each other. The entire public-school system, like the entirety of our media, supports the idea that sports and entertainment are everything; that knowledge is boring, knowledge is nothing.

The kids who go out for knowledge and education are punished by labeling them "nerds," "geeks," and "losers," or we let them win trivial trophies in spelling or debate. Then we are surprised that America is falling behind in almost every measure of success.

A recent study by the prestigious PEW research group ranked America as 24th in science and 38th in math in the world. Our history classes in high schools are a pale imitation of reality. We are the worst in education of any industrialized society on earth.

We have made sports exciting, respected, admired, and something students will want to achieve. We have made knowledge and education into a school joke, something they must do only so they can become a success in a wealthy corporation. Then we wonder why the best physicists, doctors, and scientists in America come from India, Pakistan, and China.

The media have programmed American students to want to be entertained, not educated. We are rapidly going toward the cliff to becoming a third-world nation. This is an issue we are not even discussing as we agonize over the failure of our schools in the media.

Entertainment is fun. We all need something to tickle our limbic system, to shock our amygdala, to trigger the highs and lows of emotion. The problem is that there is nothing else but National Limbic System Tickling System, in America.

I love football, at least when "my" team has a chance to win. I love a good movie; we need escapism. But the media overwhelms us with fiction and sports.

They dominate everything we see daily.

Even the "educational" channels now show more programs like "Ghost Hunters" ("Did you hear that! Wow!), and *Ancient Aliens* ("how could our ancestors have built the stone monuments without Alien Intervention?").

Sports, entertainment, and the news are so carefully crafted to get our attention, to keep our minds fixated on what will happen next, that reality cannot compete. The school system and the media have made sports and entertainment into a national obsession—more addictive than black tar heroin and just as unproductive.

Only a few decades ago, America was number one in the world in producing college graduates. Now we are twenty-second. Fully one in four American students never graduate from high school—a rate unheard of in the rest of the industrial world. We are now twentieth in producing high school graduates.

One in five Americans is "functionally illiterate," unable to do math well enough to compute the gas mileage on their oil-burning SUVs or read well enough to follow simple instructions to screw a Chinese-made oil filter onto their Korean-made car or to work the Made in Singapore remote control on their Japanese-made 3-D Plasma Limbic System Tickler.

Students come out of the vast mind-meld of a TV zombie apocalypse with little comprehension of what is important for their future and no understanding of their minds.

They come out with a burning desire to be entertained, not to learn.

Bring on the Zombie apocalypse. No more school books! Why?

Our media and schools both applaud the wonder of an athlete who can jump a sixteenth of an inch higher, run a hundredth of a second faster, or score on a long bomb in the last quarter. Is that what makes someone a hero? Sports glorification has given us a half-century of exaltation of the insignificant. Television and newspapers devote massive space to sports but very little to science and knowledge. Now we have become the Al Bundys of the civilized world, dreaming of the glory days of America.

Can we turn this around? Of course, we could. We could easily turn this around. All we would have to do is stop glorifying playing games in our schools and feeding the public intellectual tripe in the media and start making knowledge and skills exciting. All we would have to do is tell people the truth about what is important in life and what is trivia.

We spend hundreds of billions of dollars a year making movies, TV, and sports exciting, from good writing to special effects. We spend nothing making education exciting. We could start to turn it all around for less than the cost of producing a video game. We could succeed if we spent only a fraction of the amount we spend on entertainment to make education fun, exciting, and important.

Americans can choose to make this country a nation of Jonas Salk and Bill Gates and Steve Jobs instead of becoming a nation of Al Bundys and South Park.

Americans can choose to end the glorification of entertainment.

Will we?

Not a chance.

The nations that learn the lessons of motivating their youth will lord it over us in the decades to come as America continues to chase our tail in an endless glorification of schlock. Politicians and the press will continue to agonize in public over how to turn this runaway train around by doing more of the same things that have failed so completely. We will continue blogging our opinions into each other's brains as we tweet ourselves into oblivion.

Yet parents have enormous power to motivate and create success in their children as well as themselves. If enough parents learned how to do so, we might even be able to save the rest of the country so that they could continue their mass adulation of trivia and whatever tickles their limbic system. This might even keep America on top for another quarter of a century.

This is knowledge that should be shared with everyone. It should be taught in the schools—yet it is nowhere to be seen.

HOW WE COULD MAKE EDUCATION ENTERTAINING

We could assemble some of the most accomplished teachers in every subject, not the ones who published the most scientific articles, but the ones who are the best at getting ideas across, we could change our world.

We do not want the most successful professionals in the field, because they are often not good at getting ideas across to the average person. That is much like asking a teenager who has been using computers since they were ten, to describe to those of us just getting started on a computer, how to do a simple task on a computer. To them computers are like a first language, it seems so obvious that they cannot quite grasp how little we know. So, they start out with, "first, go to the settings, drop down to advanced, click on screen...." They have already lost us when they think we should know where "settings" are located.

So, find the best teachers in math, history, literature, psychology, and more, the teachers who have shown they can get ideas across, often ones who came late to the field. Get them together in a mini-Manhattan Project, with enough government funding to live in a group for a year or two. It would be incredibly cheap compared to the money we have been wasting.

Give them the task of coming up with graduated, programmed learning that goes easily from year to year.

Add to that, the best computer gaming programmers, the best video programmers, and the most creative writers. Task them with making the above data into an entertaining, exciting and thoughtful game of knowledge.

Put all of this onto DVDs that students could use to learn at their own pace. Use the same technique gamers use; they require you to kill all the bad guys in Level I before you can go on to Level II. If you fail, you have to go back and repeat Level I until you succeed and go on to the next level. Each student could learn at their own pace, and teachers would be free to work with the students who need the most help or to enthuse the students who want to learn even more.

CHAPTER 5

THE MIND OF GENIUS - LEARN FROM SUCCESS

WHAT WE CAN LEARN FROM OTHER PEOPLE'S SUCCESS

In the early years of the nineteenth century, a historian named James Mill decided he would raise his son to be the smartest man on earth. It was quite an arrogant goal to set in an era when everyone assumed that you were born to be whatever you became; good athletes were said to be "born athletes," good doctors to be "born doctors." Evil people were said to come from a "bad seed." It was all in the genes.

Today, they say it is all in our "DNA".

Mill began teaching his son Greek at the age of three. By the age of eight, he read the Iliad and began reading Euclid. By the age of twelve, he had read Virgil, the Aeneid, Horace, Lucretius, and Cicero... just for his history lessons. At the age of twelve, he had mastered elementary algebra and geometry and began differential calculus.

The son's name was John Stuart Mill. He went on to write thirteen major works, including his classic essay on democracy, *On Liberty*. This was a brilliant work, once required reading in many universities. He wrote a strikingly modern work on *The Subjugation of Women*, and his own story of the influences that shaped his life, his *Autobiography*, all of which marked him as one of the most advanced and far-thinking men of his time. Mill has been referred to as "the last man in history to know everything there was to know about everything."

John Stuart Mill noted that his father's teachings gave him a quarter of a century head start over his contemporaries. We are still far from catching up to what Mill understood about many issues, including liberty.

John Stuart Mill said that his father, *"who was not normally a patient man,"* exercised the utmost patience in teaching young Mill. Yet, for all this, Mill was unhappy for much of his life—not because of his genius, but in part, because his father had neglected his social education; and in no small part because his father was more than a little overbearing in his methods. His father had made him into a genius, but also something of a freak, in his mind, who did not fit in with the society of his time.

Years later, when Mill wrote his autobiography, his wife had to intervene to stop him from putting in all the anger he felt toward his father. Interestingly, we saw a similar reaction from Michael Jackson toward his father, as he felt he denied him his childhood. Jackson spoke of the feelings he had as he saw other boys out playing and having fun while he was forced to practice. Interestingly, too, Michael left his father out of his will, leaving all his money to others and the management to his mother.

What made John Stuart Mill and so many others into a success? And what can we learn from their parents' mistakes?

There is a vast amount of biographical data about the childhood of successful people. Some of it is as old as Goethe, whose father also raised him in the way of Mill, but who made learning a game, a fun experience. Goethe described his experience with learning as *"...a game, one I forever delighted in playing."*

Some stories are as recent as the Susedik family, who raised four children, all girls, to be in the "genius" IQ level. Their oldest daughter mastered calculus and became a sophomore in college at the age of twelve. When demonstrating how they raised their children, Ms. Susedik gave an example by reading a paragraph for them, with emotion, feeling, and enthusiasm. They wanted their children to love learning, not be forced into it.

Or of psychologists Arthur and Carolyn Staats, who raised their daughter on programmed learning. Some say that females cannot do as well as males in math, yet she was in the top one percent of the nation on academic achievement tests. Again, the emphasis was on the positive emotions of learning.

The public is deluged with stories about teaching children while still in the womb; or, at the opposite extreme, how little success comes from early learning. "Early ripe, early rot" was the dismissive comment used. What does the evidence suggest? How does such an experience affect the mind?

But before we get into the specifics of what they did right and what they did wrong, it is important to take a look at how experience affects us.

What are the keys to success in life? There is no absolute answer. Yet, one theme repeats itself often. Basic to success is a quality rarely studied in science, the origin of the emotions that set the course of our life. In the words of Charles Buxton, "Experience has shown that success is due, less to ability than to zeal..."

Parents can make a difference. They can take the frightening power of a thunderstorm and make it into something exciting. They can change the fears of childhood into something positive. That is a profound tool for helping our children, from a science not known for having any useful tools.

More than this, is it possible to apply the same positive emotions to make the task of learning life's skills exciting and exhilarating?

What if the same type of conditioning were applied, by parents or peers, to develop an interest in knowledge or books; "Ooh! Look at the pictures," or "You can learn the wisdom of history in books!" Could that develop a feeling of excitement toward learning? Would this show up in children's future success, their habits, their interests in life? That seems to be what the Susedik family used to instill a love of learning in their children.

Out of the many studies and autobiographies, I have read, I have become convinced that this positive emotional response is basic to success. It makes for confidence in children in the face of fear, leads us to carry on in the face of failure, and drives us to strive to acquire the experiences that give us the skills of living. It is basic to what our society calls "genius" and basic to our success in life.

The emotion of awe and exhilaration people describe in the presence of a thunderstorm seems identical to what Albert Einstein described in his autobiography. He refers to the "sense of wonder" that he experienced as a child when shown a compass for the first time and realized that mysterious, unseen forces could move the needle of that compass. Astronomer Carl Sagan, the creator of the famed "*Cosmos*" series, spoke of the awe he felt as a child when he realized that the "billions and billions" of stars in the night sky were also suns, like our sun. And that they, too, may have planets, even life.

It is repeated in Einstein's describing how he later came to a dramatic realization that there was an entire universe out there that *"beckoned like a liberation,"* and man's mind could understand it, at least in part., and that he, too, could come to understand how the universe works. It was not just awe and exhilaration at the thought of the universe, but awe and exhilaration at the realization that *he* could come to understand the universe.

One can almost hear Einstein's mother saying, "Oooooh, look at the needle in the compass! What is it? What makes it move?" Actually, we know absolutely nothing about what Einstein's mother did. Yet, we do know that just this type of experience has been common in successful people's childhood: Goethe, Sartre, Richard Feynman, and many others have told of its influence.

Mr. Susedik said that the thing they did differently from most parents in raising four children with "genius" level IQs was how Mrs. Susedik taught her children— with feeling. You could hear the emotion and excitement in Mrs. Sussedik's voice as she described how she taught her children to read by reading stories about their family.

This was not forced learning; it was enthusiastic learning.

QUESTIONS INSPIRE; NAMING THINGS DO NOT

Einstein said of the school system, *"It is a wonder that the (school system) has not yet totally destroyed the holy spirit of inquiry..."*

The Nobel Prize-winning physicist Richard Feynman, sometimes referred to as "the greatest mind since Einstein," tells a moving story of his father, a man who had no college experience but was widely read. When Richard was a child pulling his little wagon, he noticed that a ball inside the wagon would go along with it until the wagon stopped. Then the ball would roll toward the front of the wagon, even though the wagon had stopped. He asked his father why, and he replied that some people say the ball continues to move forward until it is stopped. "Some people call that 'inertia,' but no one really knows why." When his father would take him on walks in the woods, and they would see a bird, instead of *naming* the bird, his father would wonder what purpose its color served or what it was doing with the twigs it carried.

The significance of what his father did should not be lost. Most fathers would say, if they even knew, that "that is called inertia." As if by naming it, they had somehow explained it. But his father did not pretend to know; he left open a great mystery of nature to consider. When pointing out the bird, most fathers might name the species, if they knew, but Richard's father pointed out mysteries; "what purpose does the color of the birds?" "What makes them different", things that led to more questions, a deeper question—not a name that would give them a superficial feeling that they understood a complex subject.

The wonder, awe, and mystery of life were left an open question. That is often what we fail to do, even in college.

But this is so important an issue that it must be repeated, or it will be lost. In our schools, even our universities, when we teach students about biology, anatomy, physiology, or psychology, all we teach is nomenclature, naming things, definitions. Our subjects have no life. They are dead. What a perfect way to destroy what Einstein calls "the holy spirit of inquiry." What a perfect way to kill a child's interest in learning. Finally, our schools have found a way to do something perfectly.

I remember the thrill I had when I finally got to college; it was so unlike anything I had ever seen in high school. In high school, I learned that everything was already known; we had a name and a label for everything. What more was left to know?

Even at a university, introductory level classes were taught this way. Yet, the liberation came more from sitting in Voertmann's Book Store, sneaking a read at pop science, not so much from most classrooms. Here were great iconoclastic authors like Bertrand Russell, Clyde Kluckholn, George Gamow, Isaac Asimov, Kurt Vonnegut, Margaret Mead, and more, all of whom spoke of a reality I never knew existed before. They made learning exciting. The textbooks made it empty. The great authors talked about things unknown, an entire world never before imagined that, much as Einstein said, *"beckoned like a liberation!"*

The textbooks only talked of names and dates and gave little hint of what reality was all about. Occasionally, a great professor would give me a jolt by raising issues I had never before heard. But it was the life these authors put into their books that caught my mind. Today, few read, and you are unlikely to find what is important in life on Facebook or Twitter.

Asking questions, not naming things, is the key to inspiring children—a lesson our school system has pointedly refused to learn. Instead, schools teach only dry, sterile facts, devoid of emotion, meaning, or information.

"Columbus discovered America in 1492!" Nothing of the kind happened, says writer Kurt Vonnegut. Americans had been living here for thousands of years; *"1492 was simply the year the sea pirates began to rob them, kill them, and steal their land."* There is far more truth and more of a sense of wonder in that simple statement of Vonnegut than in anything we teach in the vapid world of the school system. School curricula is devoted to indoctrinating students with a censored view of reality, something that will offend no one, instead of teaching them to think for themselves.

Public schools are a giant baby-sitting operation. They keep kids out of trouble while parents work hard to make their employers richer. What they do teach is only designed to make us fit into the "organization man" scheme of business; be on time, be civil, obey authority, don't question, do and believe what you are told. The subjects we teach, the "back to basics" lauded by many politicians, are designed to make kids do well in the tasks required of good worker bees in the giant business beehive; reading, writing, and arithmetic. Nothing we teach is designed to benefit students as individuals, help them make it through life, or teach them to "think for themselves," as we claim to care about.

Most school systems in America are "independent." This means that local school boards, elected or appointed from the community, run them. Conservative religious groups who insist on forcing their ideology onto students have overrun many school boards. This came to a head in Texas in 2009, when the conservative majority of the State School Board, the agency that decides which school books should be used across the state, voted to only adopt textbooks that give religious "creation" stories the same status as biology's "evolution." They never thought too deeply about that one because it would have meant we would have to teach the multiple creation stories of Shiva, the Marduk story of how humans were created from the blood of Quinju, the story of Ahura Mazda of Zoroastrianism, the Gilgamesh epic of Mesopotamia, and about fifty more.

The same school board thought that there were too many non-white people included in textbooks and voted to eliminate black and Hispanic names, including Cesar Chavez, who led the civil rights movement for migrant workers, and replace him with some white guy most people had never heard of. They did have to reverse that one.

In 2018, the Texas School Board recommended that they remove Hillary Clinton and Helen Keller from history textbooks to "streamline" them. They left Billy Graham in. The spokesperson for TSB stated about Hellen Keller, "She does not best represent the concept of citizenship, and instead, military and first responders do."

And we wonder why America is now 24th in science and 38th in math in the world. The rest of the world has surpassed America in science, technology, math, and good sense. All the while, Americans eagerly await the next football game, new and exciting movies, or the latest gossip to share with our friends.

The interests that fire most of our minds begin in early adolescence, not just childhood. The same conditioned emotions can occur due to a unique experience, excitement conveyed by others, or from a chance event.

It is a mistake to assume that the early school years are the most critical. In successful people's lives, yet it is common to hear them describe critical events that occurred between ten and twelve years of age. Even their late twenties have often seen a critical event that set the course for the rest of their lives. John Stuart Mill never accomplished anything until his first publication of *On Liberty* at the age of fifty-four, only after the encouragement of his wife. But the enthusiasm and interests that begin in the early teenage years are powerful forces in determining our future.

Nobel Prize-winning mathematician-philosopher Bertrand Russell describes his early introduction to Euclidian geometry in a way that sounds more like discovering sex than math:

"I began Euclid, with my brother as my tutor, at the age of eleven. It was one of the great events of my life, as dazzling as first love. I had not imagined that there was anything as delicious in the world... From that moment until I was thirty-eight, it was my chief interest and my chief source of happiness..."

It began as an accidental moment at a critical time in his life. His parents did not plan it. It could just as easily have been football, reading, music, girls, or any of a hundred other skills that we might understand... but Euclidian geometry? How could anyone equate Euclidian geometry with first love? It all depends on the ideas that excite the mind, stimulate thought, and make us aware of what excitement there can be in understanding, knowing, and developing our skills.

Some say that it depends on a "critical period" in an individual's development. More likely, it depends on having a mind that has enough knowledge to be ready and eager to understand the importance of a new idea. The sudden realization of the far-reaching importance of an idea springs to the mind with great force. After all, we put men on the moon using Euclidian geometry, tempered with Newton's laws. Imagine the excitement the first person felt in understanding how that could be done.

We do not decide on interests that will make our future because we found something we "want" to do. These ideas do not just suddenly spring from our free choice. The ideas that fuel our curiosity may come from words spoken by a parent. We may pick up on the things that excite our friends or read an idea in a book. These things usually occur by chance. Most of the time, no one plans them; they just happen.

Yet, these chance events may set much of the course of our lives. It is common to find teenagers pick up an interest in cars, sports, or photography. They may read every book they can find, memorize meaningless statistics on football players, or develop useful skills themselves, all because of the enthusiasm or ideas they picked up at a brief moment in their life.

Should such experiences be left to the capricious accident of the moment? Could such interest be stimulated, encouraged, and nurtured?

These experiences are not what make people a success, but they provide the enthusiasm that ensures people will acquire the skills and knowledge of life. It is the emotion that ensures they will keep after it; that they will strive for that excitement of new knowledge, develop new skills, and not exhaust themselves in the face of hardship and rejection.

If psychology has a secret as dramatic as the atomic bomb was in the science of physics, then this is it. I expect that the first society that learns the secrets of motivation and education and teaches it to its future parents will have a dramatic economic, social, and scientific advantage over those societies that do not. But that is not what we do in America. Instead, we glorify sports, entertainment, celebrity, and heroes—anything but knowledge.

THE ETHICS DILEMMA:
To Choose or Not to Choose

Whenever I hear students becoming enthusiastic about making their child into the next Einstein, I feel uneasy. We have an entire industry devoted to "speak and say" toys, talking books, video games that teach algebra and trig and how to pass the SATs, and even books on talking to their child while it is in the womb.

I remind students that Einstein didn't have "speak and say" books; he didn't have a computer, and he didn't get talked to while in the womb. All he had for toys were a compass and a dreidel—a top you had to spin by hand.

I am not recommending that you mold your child into a wunderkind to make them a success at life; nothing of the sort. We have already seen that success does not necessarily bring happiness. But every parent wants to give their children a head start in life, and they are going to do so regardless of what any of us say, so we owe it to them to give them as honest as possible view of what works and what doesn't, as well as some of the pitfalls.

A mother recently groomed her daughter to become an Olympic gymnast. Her mother had even gone to the extreme of flying in a former Olympic gymnast to help train her daughter and hone her skills for the finals. Everything was right on schedule. Her daughter had a remarkable ability, agility, and skill. Everyone agreed she had what it took to be a new Olympic star, like Nadia Komenich.

But before the tryouts began, she was doing a routine when she slipped and broke her ankle. As months passed; she gained weight and grew taller. She lost her edge. By the time she had recovered from her fall, she was no longer Olympic material. Her mother's plan had failed. She had failed, through no fault of her own.

Always have a Plan B, and maybe a plan C. Do not just push on your kids the ideals of lost glory from your life. Consider some safe alternatives, like Sartre's grandfather's advice; do something else in the meantime, like teaching, medicine, or truck driving. The whole point of giving college students a "liberal arts" education is to expose them to a wide variety of information from every possible area. That is a good idea for parents, too.

Nobel Physicist Richard Feynman is a good example. His son was interested in physics; his daughter was not. That was fine with Feynman. In college, his son decided to go into philosophy, which, he says, his father probably could make no sense of, but his father did not discourage him. The son later decided on a different hard science career. His daughter went into photography. She says she always had a feeling that anything they went into was fine with her dad.

Yet, there is no doubt that parents have a dramatic impact on their children. Look at Tiger Woods, for example (before the 14-mistresses thing). There is a photo of Tiger at the age of two wearing a tiny little golfing hat with tiny little golf clubs. His dad decided early on that this was going to be his gift to his son. He groomed him from an early age, much like James Mill groomed John Stuart Mill to be a genius. But his son loved the game.

He was enthusiastic about playing. He played it on his own; no one had to force him. The key seems to be the emotion that generated enthusiasm. As a child golf prodigy, being praised in public (Garth Brooks' "roar of the Sunday crowd") would have been an added force to propel him along in life. This is the kind of emotional conditioning that works.

All of which raises some serious ethical problems. The same conditioned emotional reactions of awe and exhilaration we have just described have long been used by parents to indoctrinate their children with their own political, ideological, and religious beliefs—even their preference for a football team. Even today, it is routinely used by virtually every society on earth by pairing *their* flag, *their* national anthem, and *their* belief system with "all things good." It is that "thrill of patriotism" that runs up one's spine at the sound of the national anthem or the sight of the flag in the Olympics.

And it is that same conditioned emotional response that was used by Adolph Hitler to inspire his soldiers to greater acts of "heroism" for the Third Reich, and by Iran and Iraq and terrorists, to produce soldiers who eagerly die for God and country. This is not an emotion that is always "good." It can produce such devotion to an idea that no amount of logic or evidence can stand against it.

Politics controlled by thousands of emotionally-charged fans screaming Zeig Heil, in unthinking adulation of someone who knows how to use our emotions to control us, is not a recommended outcome.

It is not enough to create such an emotional reaction, even for the noblest of reasons, unless we also learn to understand our own minds; to understand the origin of our motives'.

We are living out our lives like monkeys in a cage. Humans know nothing about how childhood experiences shape our lives. We teach the basics of mathematics in schools, yet we teach nothing about success in life. We teach how to read and write, yet we tell them nothing about how the experiences of youth powerfully affect adult minds. We teach geography, but nothing about being capable parents.

We waste our young.

And our childhood experiences dramatically affect each of our minds. Each of us must live with the effect of our youth, yet few of us understand its force. Fewer still learn to master its force.

This is not a "how-to" book on child-rearing, although the information often applies; there are far too many variables to accurately predict the outcome of how we teach our children. It is intended as an inquiry into what is known about how childhood experiences influence and determine what we become as adults. What makes for success in life? What can we learn from the lives of those who are successful? How can we change lives for the better?

THE ORIGINS OF SUCCESS

In 1990, the man described as "the new Bach," Yefim Bronfman, was highlighted on television. In the middle of the story, they flashed a picture on screen, showing him in diapers at the age of one, standing on a piano stool with his hand on the keys.

Did he decide, at the age of one, that he was going to be a pianist?

Who did?

Did he get up on the piano stool at the age of one, by himself, to pose for the camera with his hands on the keys?

That picture tells us a great deal about who decided.

His parents did.

Yefim Bronfman described how his violinist mother and pianist father divided the children up to develop his sister into a violinist and him into a pianist.

Two television sportscasters were covering a baseball game during the summer before the 1982 World Series. As one of the players stepped up to bat, the color commentator suddenly flashed a picture on the screen of the same man when he was a child at eighteen months of age. In that picture, he wore a little baseball cap and a miniature baseball uniform and was holding an enormous baseball.

The sportscaster said, "Well, I guess that just goes to prove that great baseball players are born, not made!" He was joking, I am sure, but the same questions arise.

Did he pose for that picture himself? Did he decide on his own, at the age of eighteen months, that he wanted to put on a uniform and grow up to be a baseball player?

Not likely.

If anything, the picture of that little baseball player at eighteen months shows that great baseball players are *made*, not born.

The same story recurs countless times. Doak Walker, the All-American star football player from Southern Methodist University described how, when he was born, his father ran up and down the neighborhood, telling everyone that he had a "new All-American." Was Walker really a "born football player"? Or was he groomed early for the skills that crowned his life?

When physicist Richard Feynman was still in the womb, his father told people that if it was a boy, he was going to be a scientist.

The same questions must be asked about Mozart, who reportedly composed his first piece of music at age three. How many three-year-olds in his time were born into a family that could afford a piano? How many three-year-olds' parents would even allow their kids to touch an expensive piano? Or Johann Sebastian Bach, born of seven generations of musicians, where his father and three brothers all made their living as professional musicians.

Throughout the first half of this century, parents have often sought to mold their children's futures. They have forced their children to sit through practice sessions at a piano keyboard, all in the hope that their child might become a musical genius.

The piano keyboard has given way to the computer keyboard as millions of parents rush to give their child an advantage in the quest for the brass ring; to make their child a success in life. Computer classes have appeared throughout the country to prey on parents' fears that their child might become one of the "computer illiterates" we hear about on television.

Yet, many question these methods. Did learning to play the piano help anyone? What happened to all those budding Beethovens from previous generations?

Where are all the Leonard Bernsteins produced by so much forced labor at the piano? It is not forced labor that produces genius; it is the enthusiasm, motivation, and drive that does so. And that you cannot get by forcing kids to go to music class.

Some attempts at helping children excel were positively harmful. Many parents started their children a year early in school. Some were double promoted into a higher grade because they had already gone beyond their peers. The end result was that these children were always a year or more behind their classmates in physical development and social experience. Many could never do as well in sports or social interaction and often never succeeded as well as their classmates. This left them in the position of feeling inferior in sports and social interaction; since they were a year or more younger than their classmates, they never caught up in physical or social skills throughout the school years.

Many people believe that trying to help their children could be dangerous. Parents learned that it did not help to try to make children into a musical genius by forcing them to take piano lessons or improve their ability by double promoting them. Many began to return to the idea that talent was all in the genes; maybe the environment did not matter.

Parents and teachers went back to the comfortable role of zookeepers. Education went back to basics. The zookeeper's role is to supply the child with basic needs and let their genes fix his or her course in life. It was an easy row to hoe. If you were a parent or teacher, it absolved you of that tingle of fear that you might have done the wrong thing, done too little, might mess up, did not nag enough at the right time, or that you nagged too much. If it were all in the genes, you could do no wrong. Yet the failure of forced labor at the piano or schoolbooks ignores the most important element of all—the emotional "fire in your belly" that leads us to seek out the knowledge and experiences that lead to success.

This book is not concerned with producing great athletes or intellectuals. It is primarily concerned with how childhood experiences affect adult minds and with what makes for success in life. On the other hand, it would be difficult to avoid commenting on what the evidence suggests about what is important to success in child-rearing. There is no absolute "one way" to success... many different things are involved, and an infinite number of things that can sidetrack success along the way. Each of us has many different experiences at critical stages of our lives.

IDEAS CAN CREATE THE POTENTIAL FOR SUCCESS OR FAILURE:

On PBS television series *Frontline*, a group of American students went to Russia to speak with Russian students. Some of the Russian students expressed astonishment at a *Newsweek* article noting that one-fifth of America's entire population is functionally illiterate—unable to read a newspaper or balance a checkbook.

How could that be? They wanted to know.

WHAT IS ASTONISHING...

But that is not what is astonishing; there is something far worse. What is really astonishing is that eighty percent of the American people CAN read... and don't.

According to the American Book Sellers Convention, eighty percent of all books sold in America are bought by 15% of the population. Of all the students who successfully graduate from school, the majority will never in their lives crack another book to learn an academic subject. Some may read a newspaper or an adventure or romance novel; a few more will read a "how-to" book on auto-repair or carpentry, but very few will ever again expose themselves to anything "educational."

Our educational system has been a phenomenal success at one thing—making education unpleasant.

And Americans are surprised when Japan, Germany, China, and India surpass us in economic productivity and output of scientists, engineers, doctors, and technicians, as well as the output of quality products in every measure of education at every level.

But in the Olympics, Americans can beat Japan and Germany. We excel in playing *Minecraft, War Games,* and *Freecell* on the $1,000 computer we bought our kids to give them an edge in school.

You will find television is dominated by tennis, golf, swimming, gymnastics, basketball, soccer, auto racing, bicycling, baseball, football, and more on any given weekend in America. During the 1980 Olympic Games, I was astonished to find an hour and a half of television time devoted to the newest Olympic game... water basketball (yes, it was played *in* an Olympic-sized pool!). In 2012, it was seriously suggested that there should be an Olympic sport in pole-dancing. In 1990, ABC television devoted two and a half hours to televising the New York Marathon, where amateurs compete in a 20-mile race. Yet comparatively little time is devoted to educational facts, stories, or ideas.

Why are sports so exciting? Why does it dominate our people's time and minds? What is the origin of its success?

Imagine Doak Walker's father taking his son out to teach him his first game of football. How does he start out? Does he begin by "laying the long bomb" on his two-year-old to see if he has a future as a wide receiver? Does he criticize him for not catching a pass? Not hardly.

The secret is to ensure that the child *succeeds*. You begin with patient and gentle steps. You make certain that your son *succeeds*. You throw a ball so it *will* be caught. You follow that up with "good catch!" You comment on how well he is doing.

Success generates enthusiasm. Emotional words generate excitement.

More people go to hospital emergency rooms due to sports injuries than for drug overdoses. The pain of a back injury or pulled muscle in teenage years may plague them for the rest of their lives. Concussions can be life-changing. Dozens die each year from sports injuries, and brain concussions in football are finally becoming an issue in high school and the NFL. This was never mentioned in previous generations; only the thrill of playing was what we heard.

How could we endure this pain and still be enthusiastic about sports? Sports become exciting only because other people take the time to generate enthusiasm. Other people provide the excitement that makes it possible to endure the pain. The "roar of the Sunday crowd" provides a conditioned emotional reaction that overcomes pain and suffering. "No pain, no gain!" coaches teach our children.

We hear others cheer; we see players made into "heroes." That generates excitement and keeps up the interest. We, too, want to be a "hero," to imitate what brings the cheers and acclaim of others, and to help "our" team beat "their" team. We imagine it in our minds. We rehearse it in our fantasies.

We even cheer to see others play. Is there anything naturally exciting about watching someone toss a ball in the air? No way! It is no more naturally exciting and exhilarating than a thunderstorm. We cheer because we have been led to believe, by the cheers of others, that catching that ball is something tremendously important; it makes a player a "hero," shows skill and competence, helps "our" team win, and has been paired with the cheers and excitement of others.

Does being able to catch a ball really demonstrate skill and competence? Perhaps. But is that is going to help you in your life? Does it make a difference?

Does it save human lives the way medicine has with immunizations and antibiotics? Does it make us better people? Does it increase human understanding?

We think catching a ball is exciting... not because it is, but because our cheers make it so. We are conditioned by the emotion and enthusiasm generated by others. This is passed on from generation to generation, not by our genes, but by the emotions generated in our culture.

By such enthusiasm, society has been able to make the painful, boring, useless, and trivial games of childhood into heroic events that dominate a nation's hearts, minds, and television.

Knowledge is different. Everyone pays public homage to the idea of education, but that is not the reality of what is valued. Society does not generate enthusiasm for learning. Parents rarely take the time to make it exciting.

What happens when children come home from school with homework? Do parents try to instill a feeling of exhilaration in the prospect of learning? Not likely.

The child is often told, "Now *you* sit down and do *your* homework and *don't you* watch TV or go outside and play until *you* finish *your* homework. Do you hear me?" And we expect this to generate a nation of educated, creative adults. We expect them to learn on their own. We are surprised when they fail, turn off on school, and prefer other things.

"...People with great passions, people who accomplish great deeds, people who possess strong feelings...even people with great minds and a strong personality, rarely come out of good little boys and girls." **LEV VYGOTSKY**

Yet the whole point of our schools is to develop "good little boys and girls". The media delights in saying that the human brain has produced our ability as a species to move from living out lives of "quiet desperation" in a Hobbesian world that was "nasty, brutish, and short" to satellites in space, cracking the DNA code, and ensuring that half our children no longer die by the age of one, as they still do in primitive countries, with our development of immunization and antibiotics.

No. The vast majority of us had nothing to do with any of this. Only a tiny percentage of the population, the ones who learned the scientific method and used it, has brought us to succeSS beyond measure, while the rest of us wallow in Facebook and fiction. Entertainment rules the brain. Science does not. Science was something we were forced to swallow in school, by being fed the most boring part of science, the methods, but with no reason why they are important.

CHANGING WHAT IS VALUED IN SOCIETY

Could the same enthusiasm be generated over learning that is now cast like a wreath over sports? What would it take?

Physicist Richard Feynman describes how his father would read to him from the Encyclopedia about the Tyrannous Rex. The encyclopedia would say how tall and long it was. His father would then say, "Let's see how big that is..." and then get out a tape measure and demonstrate. His father took him on long walks, and they would stop to watch a bird or examine a flower—not just to see the pretty colors, but to examine the mystery of a bird or flower. Why does the bird preen its feathers? What is the beak used for? What purpose does the color serve? Questions create a feeling of wonder. Knowledge can be fun.

Astronomer Carl Sagan, in his monumental PBS series "Cosmos," describes his own childhood sense of awe at the realization that our sun was only one of the countless millions of stars and star systems in the universe. Indeed, Sagan's trademark later became his repeated reference to "billions and billions" of stars in the universe. Such experiences are not often planned... there is something innately exciting— awesome is a better word—about realizing that each of the stars we see in the sky is actually a sun, not unlike our own. And that these too, may have planets and intelligent life.

Such are the emotional experiences of childhood that set the adult's interests and enthusiasm for life.

What if such emotions were directed toward those things that would make for future success in life?

Would this child more eagerly seek out and learn the knowledge that produces such an emotion the way most children only seek sports or games?

A parent or teacher who takes the time to illustrate an idea with a story, make learning exciting, or excite their curiosity over the mysterious purpose of a flower's parts can provide the stimulation that sets a child's course in life. Such experiences may enhance our very enthusiasm and we may carry these memories for the rest of our lives. Yet, our natural curiosity did not inspire most of our learning experiences, but fear of failure forced them upon our minds.

Knowledge is not made interesting in our schools because the very idea behind education has been one of discipline, not enthusiasm.

Our predictions have gone awry because our very theory of learning is faulty.

It is not the students but the adults who have not behaved as we "ought."

"Before we can command nature, we must first obey her laws." Sir Francis Bacon, English Scientist

Our school systems glorify sports. We hold pep rallies to encourage athletes, train cheerleaders to cheer them, and give out letter jackets like wreaths of glory to the players. In the Olympics in Greece, they awarded a laurel wreath to the winners... twigs and leaves. Today, we award gold medals to Olympic track stars who can run a tenth of a second faster, jump a fraction of an inch higher, swim a quarter of a second faster, or even throw a javelin a bit farther. We pay them millions for product endorsements.

We have made sports rewarding. We have made education into pain. The main influence of our society and school system is not to educate our young but to glorify trivial pursuit.

This is reflected in every area of our society. It is the reason why television broadcasting recently paid the National Football League over $300 million just for the privilege of televising their games for one year, but that public broadcasting had to beg and claw and scrape for years to get a paltry $5 million to put on a quality educational program like Carl Sagan's "Cosmos."

And who are held up to the mind's eye as those to be admired by our children? Who do we see in countless television programs and news reports? Sports heroes, police stories, Hollywood stars—not knowledge, ability, or value. Entertainment sells; knowledge you cannot give away. That is the bottom line.

Decades ago, America landed men on the moon. This was an enormous scientific achievement. But who did America make into heroes? Who had "The Right Stuff"? Who were continually profiled by the media and paraded before the public? Were they the scientists who created the computers, designed the rockets, and developed the programs? No—they were hardly even noticed. The heroes were the astronauts who flew a machine that was created by the minds of others. The press profiled their hobbies, their history, their lives. The scientists received attention only when the Challenger exploded on liftoff and people wanted to know who to blame.

Thought, knowledge, and creativity are ignored. In the world of television, the closest thing to a scientist hero is Dr. House, M.D., imitating Sherlock Holmes at the age of fifty. The writing is quite strikingly original, although they unabashedly make no bones about modeling this on the original Sherlock Holmes novels. In one episode, House has to show his ID, which lists his address as 221 Baker Street. The original Sherlock Holmes lived at 221B Baker Street. But this is a soap opera, no matter how good the writing is—not an inspiration for what science is like.

And if you look at stars on television, what do you find? Most Americans have heard of Mary Tyler Moore, Ed Asner, Louie of *Taxi*, or Ted and Diane of *Cheers*. But who in America knows David Lloyd? Precious few. David Lloyd was the head writer and creative genius behind the *Mary Tyler Moore Show, Taxi*, and *Cheers*—all the very best of Hollywood. Yet he is virtually unknown.

Americans know the names of countless television stars, but how many know who Philo T. Farnsworth was? Not many. With over 200 inventions to his credit, Farnsworth was the fellow who invented the cathode ray picture tube that made television possible. He went on to develop the endoscope and the hospital incubator that saved many premature babies, as well as improving the electron microscope and more.

Farnsworth devised the first working model of television. Instead of paying him royalties, American companies fought him in court to get around his patents. Farnsworth exhausted his funds in court battles. By the time television caught on, his patents had expired. He died in 1970, reportedly embittered, depressed, and with little to show for his years of contributions.

We reward glamour and excitement in our society, we glorify trivia. The dramatically important elements that have made our technology great are ignored entirely. We *say* that we value knowledge. We *claim* that intelligence is a wonderful thing. Yet knowledge and ability are ignored by the press and often actively punished by the people. The rewards are few.

In the last 30 years, America has dropped from first place in productivity to third, after Japan and Germany. We have gone from first place in per capita income to fifth. During twelve years of the Reagan-Bush administrations, we went from 13th place in infant mortality to 23rd. As of 2010, America ranks 45th in the world in longevity, according to the World Health Organization of the United Nations; the people of 44 other nations outlive us. According to a University of Pennsylvania study, we still rank 43 in longevity in the world. We have the finest medical technology in the world and the worst medical care for our people of any civilized nation on the face of the earth. The much-feared decline in Scholastic Aptitude Test (SAT) scores is only a tiny fraction of a problem of immense magnitude.

America ranks 49th in literacy in the world. That speaks volumes. But if nobody reads, nobody knows it. If you are allowed to hear this in the news media, it is reduced to a ten-second sound bite and never heard of again.

Politicians and the press routinely trumpet the call to reform our educational system. We are told we need more discipline in our classrooms, more work from our children, and more competence in our teachers. Hardly a parent has grown up without having constantly heard about the need to develop discipline in their child. The idea being that children must be forced to do things they don't want to do to make them learn, behave, or study their school work.

Yet, forcing someone to do something only tends to work as long as you stand over them. And bribing children to do chores with money or promises only works when the bribe is large enough to offset the boredom or pain of doing the task.

So, if demands do not last and the reward of money doesn't motivate, what works? Consider this...

The problem is not to force a child to do a boring task; the problem is to make that boring task exciting, to make them want to learn.

The way many parents have done this is simple—they take the time to generate enthusiasm. They do it with their child. They make it exciting.

When parents take the time to teach knowledge and reading with their child in the same way that some spend making thunderstorms or sports exciting, when they make a boring job into something exciting and meaningful and worthwhile, and a part of helping out, when they clean up a room with the child, do homework with them and make them feel excited about helping and enthusiastic about learning, then the excitement in their tone of voice and behavior make the task exciting.

Children find it exciting to do things with their parents. Of course, that is only true if the other people make it pleasant. The reaction will be quite different if they nag and criticize the child. The result will hardly be enthusiastic if they only tolerate the child.

Imitation can be exciting. Working with mom and dad can be stimulating. The feeling that you are helping out can be exhilarating—if the parents make it so. Being nagged and criticized is no fun. Having to do it all by yourself because you are "supposed to" is a terrible punishment. There is no reinforcement of social interaction, no excitement of talking, seeing, and learning along the way; there is no praise and no feeling of the value of what they do; much like school.

If the same parents would take the time, beginning at an early age, to go through the motions with the child, to make them feel a part of the family and tell them "what a help you have been," then that could help establish even the most unpleasant of tasks as being all things positive.

Philosopher-historian George Santayana said, "*We only really become good as adults at those things at which we have played as children.*"

Fathers such as Richard Feynman's took the time to make learning exciting. It only took a few minutes more. In primitive tribes such as the Dani of New Guinea, children play at the jobs they will do as adults, imitating the tasks they see the adults do. They play with little saw grass spears at hunting and at war, and they follow along after planting yams, imitating what they see. Play is the basis of success because it is exciting; it provides the spice of life, the emotions that carry us through.

"I seem to have been only like a boy playing on the seashore, and diverting myself in now and then finding a smoother pebble or a prettier shell than ordinary, whilst the great oceans of truth lay all undiscovered before me."

Sir Isaac Newton

Even work does not have to be drudgery; it can give the child a good emotional feeling—of being a part of an experience, pride in making a contribution, of doing something worthwhile, and happiness in helping out their parents. That is an emotion far removed from having to do something "because I said so" or "because it is your job." But to do this means we must let them know we appreciate their help—do it with them; do not make it into something they are forced to do alone.

Once this feeling has become sufficiently well established, it tends to persist on its own. Parents can back off and let them have their own direction. This feeling continues to fuel their curiosity and generates the drive that makes for lifetime interest.

Parents often have as little enthusiasm for working with their children as their children do for doing their homework. Parents fail to do it right because no one ever took to time to teach *them* how to do it right, to show them what can be accomplished, to generate in them the enthusiasm that can be felt when they know they can make a difference.

Of course, experienced parents quickly point out that even the best advice often fails; fine-sounding advice in books often seems useless in the real world. What works with an eight-year-old may not work so well with a teenager. Yet, this is largely true because there are so many competing forces for a child's attention. The fun of playing, the social excitement of talking or arguing, plus video games, movies, and going to the mall with friends all catch a teenager's mind, where each new stimulus can easily overwhelm the "fun" of doing homework or helping out around the house.

Competing forces interfere with even the best-laid plans. Reading will not easily compete with the excitement of television or the fun of playing with others. Avoid the competition. Do it when there is no distraction.

It may only take a few minutes extra to add what is needed to make it exciting and to do it with the child enough times until it becomes established.

None of this is new to psychology; what is true for children is true for adults. If you study personnel management, you will learn the same techniques. How would you feel if your boss were to demand that you do your job or berated you for not doing a better job? How would you feel if he gave you a job to do and then ignored your contribution? Most of us work better when we believe that others appreciate what we do, when we can share other people's enthusiasm, and when we feel we are making a contribution. Yet, we seem to feel that children should be treated differently than adults.

Many feel that children should be punished for misbehaving, forced to conform to rules, or ignored for their small accomplishments. Adults might resent this. Why would we expect children to feel differently from us?

"Making it exciting" is the best way to establish the skills they need in life, though it is easier to start this with a six-year-old than with a twelve-year-old. However, it is never too late—even adults appreciate being appreciated and prefer fun group activities to things they have to do alone. The techniques are as basic to adult, spouse, or employee psychology as they are to children.

It is not too late just because you did not talk to your child while they were in the womb or teach him or her Greek at the age of three. Indeed, there is no evidence that talking to your child in the womb or teaching Greek at the age of three has any value.

THE NEGATIVE FORCE:

One of the lessons I learned many years ago when I taught child psychology courses to students, mostly parents taking night classes, is still worth noting. The kind of people who take courses in child psychology or who might read this book are not the kind of people who error by doing too little. They are the kind who try too hard, do too much, and are too nice. Even being too nice may be a problem if it fails to prepare your children for the reality of what people are like.

The dangers are considerable. People often overdo their role, expect too much, praise too often, enthuse too much, or over-react to little things. It all works on the general principle of "more of a good thing is even better." Not so—it creates the "grandmother syndrome." Grandmothers may praise every scribbling as a work of art. The child shows the same scribbling to a kid next door, and they trash it. Children quickly learn that others in the real world do not share this value. Too much praise dilutes its value; a light touch may be best in all things.

I often recommend that parents spend fifteen minutes a day in a face-to-face conversation with their kids, getting them started on their homework, and enthusing over something important. But when I look back on my own life, I cannot remember spending fifteen minutes a day with my parents, even though my mother went out of her way to teach me reading, writing, and arithmetic before I ever went to first grade, and a love of reading. After that, not so much.

A light touch may be best.

Yet the few times I do remember her spending time talking about something had a profound influence on me, to this day. Parents who think they have no influence on their children may believe that they have no success because it may take decades before that influence comes to the surface; in their choice of occupation, their strategies of living, ethical ideas, and their childrearing practices, for better or worse.

THE PITCHER OF WARM SPIT:
Too Much, Too Soon, Too Often:

If positive emotions can inspire, then unpleasant emotions may destroy the interest the child has in learning. If not doing enough for their children is a problem for some, then doing too much is a problem for others.

Over-enthusiastic parents, with the very best of intentions, may produce just the opposite emotional response. Forcing learning on a child's mind can be a great mistake.

A young mother, a student of mine, shared that she had started to teach her daughter to read. Her daughter kept making mistakes, and she kept correcting those mistakes. Pretty soon, her daughter said, "I don't wanna do this anymore. It's no fun."

It is more important to make it fun than to force them to learn. Trial and error are basic to all learning—even in adults. The mistakes will naturally disappear as they gain more experience. The fun is harder to get back.

Albert Einstein once described his own experience with forced learning in the university system and how he was unable to think or work for a year after taking his exams:

> *"It is nothing short of a miracle that the modern methods of instruction have not yet entirely strangled the holy curiosity of inquiry... It is a very grave mistake to think that the enjoyment of seeing and searching can be promoted by means of coercion and a sense of duty".*

I can still remember the trouble I had when I was forced to learn a foreign language in college or read journal articles for my thesis. Again and again, my eyes went over the words, and again and again, I learned nothing. Facts in which I am interested I could learn with little effort, but when I am forced to learn something for which I saw no purpose, it became dull, boring, and useless. It was painful to be required to learn a language I would never use or to force myself to go through countless journal articles for my thesis that said nothing important. It is not the problem of motivation per se—I could continually force myself to keep reading the material—but the mind itself is repelled.

Once I finished my thesis, it was over a year before I could force myself to read a book, or even a newspaper.

The only example to explain how I felt is the response from Vice President John Nance Garner when asked what it was like being Vice President. Vice Presidents had no power. He replied, *"It's like being forced to drink a picture of warm spit."* The only comparison I can think of is that trying to learn something this way is like trying to force one's self to drink John Nance Garner's famous *"pitcher of warm spit."*

> *"The first aim of a good college is not to teach books, but the meaning and purpose of life. Hard study and the learning of books are only a means to this end. We develop power and courage and determination and we go out to achieve Truth, Wisdom and Justice, if we do not come to this, the cost of schooling is wasted."* JOHN B. WATSON

DYSLEXIA: And the Pitcher of Warm Spit

What if that same pitcher were forced on a child who was first beginning to read? What if he or she were force-fed reading and writing? How would the child react? It is possible they might emotionally revolt against learning this subject as completely as my mind revolted against those words I kept reading over and over again.

It should surprise no one that once people leave school, most will never again show an interest in reading or learning. It should also surprise no one that so many drop out. If adults' minds react so badly to such experiences, what would happen to a child's mind?

We hear students often say of math, "Oh, I just can't do math. My brain can't do it." No. Anyone can do math, if it is taught right. Math has to be taught in segments, beginning with the 1,2,3's before you learn addition, and then subtraction, before you learn division, then multiplication...

If we miss any single step along the way, we fall further and further behind. If we are not taught the value of math, what math allows us to understand - that otherwise makes no sense, how math allows for contrast and comparison in science, then why would we want to learn math?

The same can apply to reading. If we start school behind in reading, while others are doing well, this is the beginning of the problem. What does the teacher do? Teachers start by going around the room, from student to student, having each one read a sentence. When you get to the kid who, for whatever reason, never learned the basics of reading, he stammers and stutters, and flubs it.

The embarrassment is palpable. In front of everyone.

Is it any surprise that we may hear kids say, "I don't want that book learning, I would rather do sports." So, they try to hurry through reading in class, which means they will make more and more mistakes, and get more and more discouraged.

If a child with no preparation for learning were to be sent into our public schools and failed to learn in the first few years, then this might set up a long-lasting emotional reaction to learning. Each failure feeds the negative emotion. The next time they try to rush through the reading to get it over with.

A parent who was overly concerned with trying to give their child a head-start might try to force-feed reading and writing to their child. In effect, too much, too soon, too often. Enough such experiences could produce an emotional reaction that could look like a "learning disability." I do not doubt that some people have a real learning disability, such as dyslexia. Yet, it may be significant that so many children diagnosed as "dyslexic" or a "reading disability" have parents who are highly educated and perhaps highly motivated to force their children to be a success.

Before anyone is diagnoses as having a "reading disability" and assumed to be unable to learn, the very first consideration should be to the likelihood that motivation, not brain function, is the reason for the problem. Interestingly, the technique used by speech therapists would seem to work regardless of the cause but it suggests a motivational reason.

What speech therapists do is to train the individual to "Sloooow dooown their speech. That is the type of treatment that could work because the problems results from the student trying to rush through having to read, to "get it over with". And that is a major cause for dyslexia in the first place.

"A DOZEN HEALTHY INFANTS..."

"Give me a dozen healthy infants, well formed, and I guarantee to take any one at random and make of him anything you wish, doctor, lawyer, merchant, chief, regardless of his talents, penchants, abilities, or the race of his ancestors. I am going beyond my data and I know it, but others have been going much farther on considerably less evidence." JOHN B. WATSON

Watson has been widely criticized for going beyond what we know, yet he was far more correct than his critics; *"regardless... of the race of his ancestors"* was far beyond his time. When he made that statement in the 1930s, America was a vastly more racist country than today. People believed that your biology was destiny; no one would have believed that a Black American could grow up to be President. The same people who criticized Watson would never have thought it possible. Part of the criticism may have been reasonable, yet Watson has proven more correct than his critics.

Watson did not know then all the influences that change our direction in life; that make us who we are. The vast number of experiences we each grow up with can change our potential, for better or worse.

Studies carried out by sociologists show that the average American father spends about five minutes a day in actual face-to-face conversation with his children. Much of that time is spent at the dinner table telling them what they should or should not do. It is not productive or useful.

Parents who spend only a few minutes a day, as Richard Feynman's father did, getting their children excited about reading a good book, asking questions and going over their homework with them, or enthusing over a flower, can make an enormous difference.

When successful people describe their own parents' best qualities, they often talk about the emotional feelings they had while learning. The best of parents generate enthusiasm. They create excitement, tell stories, and ask questions. They let the child show off what he or she has learned. They don't overpower them with how much the parent knows. They don't preach or lecture. Instead, they create wonder, tell stories, question, and enthuse. All this is what the school system cannot do, not the one we have now.

One multibillionaire spent time traveling through the country to high schools to give a keynote address. One of his bits of advice is...

"Be kind to the nerds because one day you will be working for one." BILL GATES

CHAPTER 6

CHILDREARING OPTIONS YOU NEVER KNEW

OPTIONS OUR PARENTS NEVER HEARD OF ...

"I remember the rage I used to feel whenever an experiment went awry. I used to shout at my subjects, 'Behave damn you. Behave as you ought.' Then I realized that they always behaved as they ought; it was I who was wrong. I had made a bad prediction."

T. E. FRAZIER IN B. F. SKINNERS *Walden Two*

The same is true of parents and our educational system. If students do not "behave as they ought," it is because *we* have made a bad prediction; we have failed to understand what is important in stimulating them to seek the knowledge and skills that are important to living. We have exalted entertainment over education.

Sports have been made exciting. Intellectual activities, knowledge, reading, writing, arithmetic, or studying music and computers have been made into a painful task that must be endured or forced on the child "for his own good" so that he can "make something of himself."

"A father would do well, as his son grows up, and is capable of it, to talk familiarly with him; nay, ask his advice, and consult with him about those things wherein he has any knowledge or understanding. By this, the father will gain two things, both of great moment. The sooner you treat him as a man, the sooner he will begin to be one; and if you admit him into serious discourses sometimes with you, you will insensibly raise his mind above the usual amusements of youth, and those trifling occupations which it is commonly wasted in." John Locke

MORE CONTROVERSIAL THAN SEX EDUCATION...

More important still is the need for our schools to teach children to become decent, effective parents with a few parenting skills. Knowing this is even more important than learning the "back to basics" of reading, writing, and arithmetic. And we can only learn this by becoming aware of how our own minds work and by educating our youth to understand these skills.

One thing that has prevented all progress in educating our children is the rise of the Independent School System. Independent simply means that local school boards run them, who often use their power to censor ideas they do not want their children to know. Personal politics, local religion, and individual beliefs have all conspired to create the worst school system in the civilized world, with the U.S. scoring 24th in science and 38th in math. Undoubtedly, we would score a zero at "teach the young what they need to know to do well in life" or how to be a better parent. Parents would be outraged if schools taught that there are more effective ways of working with children than they know.

"You told my kids I am a lousy parent!" is the mantra teachers would hear coming from parents across the country.

Most people somehow realize that the American family is the basis of building a better society. We cannot change rapid increases in crime, homelessness, and poverty by legislating social change. In the last 60 years of social welfare, giving away money to the poor did not help. In the 60 years since Richard Nixon first started to "get tough on crime," building new prisons and putting more people in jail has not helped. America's prison population is now six to seven times more than any other nation on earth. Our population's incarceration rate is exponential, like Al Gore riding up a lift to show how high the exponential curve in carbon dioxide was that leads to global warming. We have been polluting our nation as surely as we have been polluting our environment.

If these problems can be solved, then it must begin in the schools. And the only effective way to do this is to teach young people to grow up, to become caring, capable parents with a few childrearing skills, and to give them the tools to be good parents, just as we give them the tools to read and write, or to add and subtract. This is essential for our schools to do; it should not be an elective. Yet, we do nothing. Most of the industrialized world has already taken steps to look at honest solutions for these problems. Most require some degree of parenting skills to be available in the schools. We have not.

Why do we not teach children to become effective parents? Unfortunately, it is even more controversial than sex education or prayer in schools. Millions of parents would be outraged at the hint that they might be doing something wrong; that their child might go to school to learn something that would make them seem less than adequate. The banner of their ego would be trashed.

Yet, we must start somewhere.

Aside from this, the answers to how to become a good parent are not always clear. Successful people come from all kinds of backgrounds, not just from the "best" of parents when even they often have problems. Pain and suffering are common in the most successful people. So, the solution is not easy. Yet it must come. Our failure to teach children to become capable parents shows up in every crime statistic, family living in poverty, and psychiatric hospitals, and countless day-to-day problems.

Children are faced with enormous problems simply in getting through a day in the school system. Academic subjects are the least important problems they face. They must learn to cope with social interactions, dating and sex, gangs in the schoolyard, bullies on the playground, and others who plant emotionally charged ideas in their minds. Being a success in life is not just a matter of intellectual ability; it also includes coping and social skills, practical knowledge, and emotional strength.

In this world, each of us must succeed in our own special niche in life. Making the most of what you have is being successful. Creating a satisfying life is being successful. Avoiding the mistakes of others or of our past is being a success. The goal of life is not to imitate famous people's lives but to improve our own. Fame and accolades of the public are not what make life a success. "All glory is fleeting," said the sage to the glorified athletes, the Olympians of Homer. Instead, the goal should be to improve our children's lives and the ability of each of us to make our lives a success.

THE MOST DANGEROUS MYTH IN AMERICA

Every year in America at least a few of the hundreds of cases of "Shaken Baby Syndrome" make the news. These are cases where a parent shakes a baby so hard that it may snap their neck or damage internal organs. When asked why, why did you do it? Most reply with the same reason, "He wouldn't stop crying. I had to teach him."

In one sad case a woman's boyfriend was left to care for a baby while the mother went to work. The baby started crying, while the boyfriend was watching the World Soccer Championship on TV. Yelling at the baby to "stop crying" did not work. He would not stop crying. Spanking made it worse. So, the father shook the baby hard to, "teach him a lesson." The baby died. The boyfriend stuck a screw down the baby's throat to pretend the baby died from swallowing a screw. It did not convince anyone.

The "Teach him a lesson." Myth is firmly embedded in the national psyche. Kids hear it growing up, "if he doesn't do right, hit him, *that will teach him.*"

You hear the same excuse in cases of child abuse, spousal abuse, and even police abuse. And if the child makes the same mistake again, which they will, they may hit them harder the next time. And again. This may snowball into an increasingly vicious behavior.

The protests over the murder of George Floyd by a police officer who was filmed with his knee on his neck for eight-and one-half minutes, far beyond any need to subdue him. Bystanders pleaded with the officer to let him up. Floyd died.

It may not be possible to say what the officer's motivation was in a specific case. But this is the kind of behavior commonly found in police abuse, where officers feel they are "doing their job" to "teach them a lesson".

"…if the only tool you have is a hammer, everything looks like a nail."
Psychologist Abe Maslow

When the only childrearing tool parents know is the "teach 'em a lesson" cliché, everything calls for punishment.

How can they possibly change their behavior? We have not taught any alternative methods in our culture. Our schools avoid the most important information in favor of "back to basics".

Physical Punishment is not as Effective as Parent Think

Psychologists often tell parents, don't hit, do a time-out. But they never explain how difficult it is to do a time-out. So, parents try time-out and find it does not work. This failure leads them back to the old ways.

Parents believe punishment is effective largely because if a child is doing something "wrong", and they hit him, it stops the ongoing behavior immediately. That convinces the parent that punishment works. It may teach them to be afraid of you. But does that really teach what you want to teach? We rarely even think about what we are trying to teach.

Psychologists fail to explain why physical punishment is not effective. Until we explain that, we cannot expect anyone to understand what psychologists are talking about.

First, when psychologists have asked kids if they have been spanked, they can certainly tell you the answer. But when you ask them *why* they were spanked, they say, "*because I was naughty*". Then when you ask them what they did that was naughty they say, "*I dunno.*"

When children are afraid of getting a spanking, or when they have the pain of a spanking, that fear and pain are all their minds can focus on. The automatic "fight or flight" mechanism in the Limbic System, grabs all of their attention. Fear takes away every other memory and leaves only the spanking.

When one spanking does not work, parents try more. But the "hit 'em, that will teach them" does not work when their brain is only focused on the fear or the pain.

Instead, calmly explain to them what they did that was wrong, without raising your voice. Don't expect that one explanation is enough, but it is far more likely to be remembered.

There is an excellent example of this with a five-year-old. Every time they got into the car to go somewhere; he would suddenly have to pee. So, everything would stop until they took him back inside to pee.

Clearly, this was irritating to his parents, but his father calmly explained to him, before they got into the car next time, that they were going to have a new rule. Now, every time before they get into the car, he would have to pee first. He asked his son to repeat what he had just said, and he did.

Now the next time they start to leave he will immediately bolt for the car. But that is an opportunity to say, "now what is the new rule before you get into the car?" and he may say, "Oh, yeah". This gives you a chance to repeat the learning experience.

It took Pavlov six trials, on the average, to get his dogs to simply salivate to the sound of a bell, don't expect that kids are going to learn just by being told once. Even adults rarely learn it right the first time.

Think About What You Are Trying to Teach

That cliché of "teach 'em a lesson" is so deeply embedded in our culture that parents often do not even think about what they are trying to teach.

If you ask people "How do you teach a puppy not to pee in the house?" The vast majority of people will say something to the effect of, "…hit them on the nose with a rolled-up newspaper", or "Rub their noses in it."

But what is it you want the puppy to do? If you hit him with a rolled-up newspaper it will make him afraid of you. If you yell at him he may be frightened. He may lower his head and look ashamed. But as soon as you leave the room he will go right back to peeing.

Think about what you are teaching. What do you really want him to do? You really want him to go outside to pee. How is hitting him with a newspaper going to teach him that?

I have raised puppies all my life and it is amazingly easy to teach them what you want to do, if you do it as soon as you get the puppy home. That first day, when everyone is paying attention to the puppy is the time to teach. After a few weeks, no one will be paying him any attention.

Wait until you see he is going to pee. Some will telegraph what they are going to do by turning around and around before they pee. Others will just cut loose. Whatever they do that is the time to pick him up, take him outside, put him on the grass, and pet him up.

To my amazement, it only takes about two times, ant the next time they want to pee they will go scratch on the door to get out.

But there is a serious caveat here. If the first time they pee indoors it is on the rug, inside a closet, for the rest of their life they will consider that their toilet. Catch them the first time, the primacy effect rules.

There are no guarantees that this will work at every age, certainly when they are small, they routinely just pee in their nest. Usually, you will have to catch them sometime between two and four months of age when you bring them in the house the first time.

With children, the problems are the same. In one example a mother gave her son a set of crayons, a common gift for a child. He immediately started scribbling on walls, floors, sofas. It kept her busy just trying to stop him from making a mess until she had to take it away. Yet what else can you do with crayons? Shouldn't parents know better?

But kids are too young to argue with their parent's logic. Yet. All too soon they will learn to argue.

THE HIERARCHY OF WORKING WITH CLIENTS

Before anyone is allowed to go to work with the Department of Mental Health and Mental Retardation, every future employee, even the secretaries, are required to take a course in PMAB; the Positive Management of Aggressive Behavior. It is the kind of course that should be required of parents, teachers, and police.

These are the kinds of methods that work for everyone, even spouses.

REDIRECTION: One of the simplest, and most effective is something mothers almost automatically use with a baby. If the baby wants something it can't have, like a shiny piece of broken glass or food on the floor, it may throw a temper tantrum if the baby can't have it.

The most effective way to deal with this is to get the baby's attention focused on something else. "Ooooh, look at the pretty picture! What is that? What color is that? Can you draw this picture? It doesn't matter what the words are, the emotion in the tone of voice, pointing at something, all of this is likely to get his attention focused on something else.

It may take a while to succeed, but it is generally very effective. Sometimes you will just have to move the baby to another room and hold on until the baby gets over the tantrum. Nothing works all the time, but women get good enough at redirection they often use it on their husbands.

REDIRECTION:
FOR MENTAL HEALTH AND POLICE PROFESSIONALS

Suppose you are working with mental health patients in an outpatient clinic, or you are a police officer called out to deal with a mental health problem. A client is threatening you with a knife. What do you do?

In an outpatient clinic, you might threaten the client by calling the police. But if he is already angry that might make it worse.

The same with the police, if they threaten the client it may make it worse. Police are routinely taught that they have an absolute right to kill someone if they think their life is threatened. Too often they tend to go to that first.

Instead, the most effective way is to start with, "What happened? What made you so upset? Tell me about it so I can help. Did someone hurt you? "What is it?"

Keep asking questions. Studies have shown that the longer you keep them talking, the greater the chance of a peaceful outcome. But this idea has to be taught over and over. It is very hard to get across after police have already been taught, over and over, that they have a right to kill someone if they feel threatened.

Very often people are angry and upset because no one will listen to their side of an issue. Just listening, non-judgmentally, can often defuse the problem. Not always, sometimes it will take extra help from the police or mental health professionals. But if someone resorts to threatening the individual or tasing them to make them comply, that may trigger a more dangerous escalation.

The FBI hostage negotiating team is trained to respond much the same way. The longer you can keep them talking, the better the chance of a good outcome. This is the kind of training that should be universal in police agencies; instead, they are taught to shoot center mass, not to shoot to wound, and keep shooting until the "threat is eliminated".

Often, they are told horror stories of a cop who shot a "bad" guy, thought he was down, and then the bad guy shot the cop. Those kinds of events are extremely rare, but when rookies are taught such things it increases the odds, they will empty their weapon on a subject. They are told to "keep shooting until the threat is eliminated."

In one case in Florida, police were told a subject had attempted to kill a police officer. Four squad cars cornered the car and fired over one hundred shots into the car, killing the man and his girlfriend, risking the lives of bystanders in the process. It turned out they had the wrong car, the two people they killed were innocent. It was like nothing so much as a feeding frenzy. One hundred shots to kill two innocent people who had threatened no one. But they were not fired, not even for being lousy shots.

EXTINCTION PROCEDURE

There is a remarkable video of a two-year-old who is trying to get his mother's attention. He follows her around but when she ignores him, he falls to the floor and starts crying. The mother ignores him and walks past him. He gets up, follows her again, sees her in the kitchen, and falls down, and starts crying. Again, she ignores him, walks right by him, even the dog ignores him. He gets up follows along, and again falls down and starts crying.

What is going on here? At first the two-year-old has learned that he can get his mother to come and play with him if he just cries. But when the mother is busy, she may ignore him. Eventually, he learns, probably by accident, that if he falls down and cries, the mother will rush over and comfort him and play with him.

Eventually, every time his mother is busy, he pretends to fall down and cry, just to get his mother to stop what she is doing and give him attention. Now this behavior is so reinforced by getting him what he wants, that it is incredibly resistant to extinction.

The mother is doing an extinction procedure by ignoring the inappropriate behavior. Yet it is incredibly persistent. When the behavior is this intently programmed, it generally takes three days of extinction by ignoring the behavior before you begin to see an effect. Many parents are not so patient.

But one thing that is essential in an extinction procedure, in addition to ignoring the inappropriate behavior is to find a way to reward a more appropriate behavior. This makes the technique more effective.

To reward a more appropriate behavior, the mother might say, "Come, help mother, sweep the floor." Or "Come, help mother, dust the furniture." And give him some attention for doing something useful. Then, gradually back off in the amount of attention you give him and only praise him when he finishes. Or, have her color in a coloring book and then show mommy when she finishes.

DO A TIME-OUT?

One of the biggest mistakes psychologists make is to just tell parents to do a "time-out" if the child is bad. But the parents try it and it never works!

The reason it does not work is that psychologists do a very poor job of explaining just how difficult it is to do an effective time-out.

FIRST: Before you ever start to use a time-out, sit down with the children when they are NOT in trouble, and explain the rules. Write them down if the child is old enough to read. "No hitting, no biting, no backtalking, no name-calling, etc." Explain this now, as well as the consequences, so you can remind them of the rules when they, inevitably, break them.

SECOND: When they break the rules, **verbally warn** them that if they do it again, they will have to go into time-out. Remember, the whole reason for the warning is to teach them, not to just punish them. If the problem is serious, as in hitting or biting, you may need to go directly to a time-out and explain why.

THIRD: When they inevitably do it again, take them to the time-out place, where you can still keep an eye on them, but still apart from the offense. Remind them once about the rules, otherwise do not talk or argue during the entire time-out.

FOURTH: Keep them in time-out, with no toys, no iPhone, no entertainment, for one minute for every year of age.

FIFTH: Before taking them out of time-out, ask them what it was they did that was wrong. Remember, the whole point is to teach, not to punish. Indeed, the point of the time-out itself is not to punish but to stop the ongoing behavior immediately, and help them to learn to stop their own behavior.

Keep in mind that some kids such as those with ADHD may be too impulsive for this to work effectively. Still, it is better than punishment.

Never forget that, as adults, we do not learn new ways of behaving easily. It is hard for us to lose weight, or learn a new language, or change our bad habits. We should not be too ready to expect too much of our children. Actually, it is easier to change a child's behavior than it is our own. Once any behavior is deeply ingrained, it takes a great deal more to change it. But we would not think highly of the idea that adults should be punished for overeating, or failing to learn.

DON'T LECTURE, EXPLAIN.   Before they ever break the rules, such as staying out way too late without calling, take the time to explain why parents get so upset if they stay out too late. If there is a time when they are expected to be home, and they are an hour late, every parent freaks out. Why? Because they think the very worst has happened. They think the kids are "dead in a ditch" or they will be a headline fo a missing person in tomorrow's newspaper. Irma Bombeck, the news columnist, said that every time she hears a siren, her heart jumps. She thinks, where are my kids? Has something happened to my kids?

Give them that bit of understanding and it will make life easier for both of you.

DO NOT WAIT FOR A CHILD OR TEEN TO BREAK A RULE AND THEN COME DOWN ON THEM. EXPLAIN WHAT THEY NEED TO KNOW BEFORE A PROBLEM HAPPENS WITHOUT GETTING ANGRY. HAVE A CALM DISCUSSION WITH THEM ABOUT IT BEFORE THEY HAVE A PROBLEM.

LISTEN, DON'T LECTURE: Who do your teens go to when they have a serious question about sex, drugs, life?

As they get older, it is important to let them know they can talk to you about anything, sex, drugs, rock, and roll, without you freaking out and giving them a lecture on what they should think. they must be able to trust you well enough to know you will be there to help, not to lecture. if they do not feel they can talk to you about their boyfriend or girlfriend, or any serious subject, they will not have a safe person to talk to when they encounter serious problems in their life. Instead, the only ones they have to talk to about sex, dating, drugs, social situations and more will be their own friends, few of whom know very much about life, because we have failed to teach what they need to know in our schools.

If you want to get ideas or values across, give them examples, tell them stories, like the stories at the beginning of this book about the kind of problems real people have that no one ever talks about.

Don't wait for them to ask. Give them the knowledge with honest, real-life examples. No parent can possibly know all of the problems they will encounter, the best anyone can do is give them a good basis in understanding reality.

Eventually, all children will become independent adults.   That is the goal of being a parent.

"When I was seventeen I thought my old man was the stupidest person in the world. When I got to be twenty-seven, I was amazed at how much the old man had learned in the last ten years." Mark Twain

CHAPTER 7

THE PROPHYLAXIS PROJECT

KNOWLEDGE AND THE ART OF PREVENTING PROBLEMS

MOST DISEASE IS PSYCHOLOGICAL IN ORIGIN;

SAYS HARVARD MEDICAL SCHOOL

Harvard Medical School published a report saying that the majority of diseases that send ordinary people to doctors or hospitals are psychogenic in origin. What?! Psychology causes disease?! When I first heard that, I thought it was nonsense. Cancer and heart disease are not caused by psychology. Stress is a factor in heart disease, but it is only one of many. The idea seemed incredible.

However, their reasoning was impeccable—smoking cigarettes causes 85 % of all lung cancer. What is the number one cause of cigarette smoking? Not our genes or our biology, but our environment. The single biggest factor in determining whether you will smoke, drink, or use drugs is whether your friends (peers) smoke, drink, or use drugs. Nothing else even comes close, not your genes, not your DNA, or your biochemistry—the peer group rules.

Type 2 Diabetes, high blood pressure, heart disease and strokes are related to the food we eat. What determines which foods we eat? Habits learned early from our mothers, or huge amounts of sugar and salt added to our foods by companies who know how to make us want more.

Once you have it, treating lung cancer is overwhelmingly invasive and far less successful. Preventing it by teaching young people about peer pressure and the dangers of smoking would be far more effective.

It is hard to get this information across to young people because they may smoke, yet they feel no ill effects from it. It takes decades for the ill effects of emphysema, lung cancer, and heart disease to occur. According to the report from the Surgeon General of the United States, some 350,000 Americans die just from smoking every year. That is far more than the 60,000 estimated to die from the opioid epidemic that now gets attention from politicians and the press. The media's preoccupation with the latest sensationalism has sucked all the oxygen out of reality. So how do you get the information across? Lecturing them not to smoke is worthless; they ignore it totally. Even adults ignore being lectured.

The most important way is by a technique used in psychology, counter conditioning. Simply show them pictures of an autopsy of a fifty-year-old healthy lung; it is pink and solid. Then show them a picture of a fifty-year-old lung of a smoker; it is black and laced with holes where the lung tissue is eaten away. You do not have to lie to them; you do not even have to tell them the truth; you *do* have to *show* them, visually. Already we have reduced smoking by one-third, just through education and the simple technique of counter conditioning. Putting heavy taxes on cigarettes is also is a major player.

What is true of smoking is also true of drinking, using drugs, sexually transmitted diseases, crime, and automobile accidents. What?! Automobile accidents? Yes, and much more.

A study from the Harvard School of Public Health by Ding, Mozaffarian, Taylor, Rehm, Murray, and Ezzati in 2009 of the preventable causes of death in the United States found that the two greatest causes of death in America are high blood pressure and tobacco (related to diet and psychological habits we pick up in our family and environment). Over 467,000 Americans die each year just from those two killers. This is followed by other psychological/cultural killers, such as being overweight (216,000) and physical inactivity, which kills an estimated 191,000 each year. All of these are due to environmental factors, psychology, learned eating habits, fast-food diets, and more. Type 2 diabetes is related to the foods we *learn* to love. Alcohol, like smoking, is overwhelmingly related to whether our parents or peers drink. These are the forces in the environment that kill most Americans.

What determines if you will become an alcoholic? The geneticist who discovered the "gene" for alcoholism stated that it could account for 12% of those who become alcoholics in an interview on the news show *Nightline*. Yet immediately, the newsman turned to the camera and said, "There you have it, alcoholism is caused by genes." No, that was not what he said.

Since then, people went into overdrive with wild estimates, making their way into the media and even into textbooks, claiming that 50%, 60% or more of alcoholics were that way because of their genes. Recent studies suggest that much of the early evidence may be faulty because it was based on post-mortem evidence from people who had died of cirrhosis of the liver. Alcohol was once the number one cause of cirrhosis. Today, the number one cause of cirrhosis is Acetaminophen; as in Tylenol.

Yet, we have known for a very long time that the single greatest factor in determining whether you will smoke, drink, or use drugs is whether the friends you run with smoke, drink, or use drugs. Nothing else even comes close. This is not to deny that there may be genetic components in alcoholism and other psychological components as well, only to try to put the perspective where it belongs—on the forces in the environment that make it happen.

But the human brain is an emotion-driven machine; reason and logic fail when they come up against emotion. Simply telling people to put reality in perspective does no good; it is essential to begin to protect them before the problems arise by educating them to understand the power of the peer group, the problems of emotional conditioning by society.

If we want to help ourselves and our children, we must start with preventing those problems in the first place by correcting the forces in the environment that create the problems of living, by learning to understand how our mind works. You can't change your DNA; you can only change our education, the peer group, and society.

Unfortunately, we have a media that is so enthralled by the hi-tech illusion, the magical, mystical belief that biology, DNA, biochemicals in the brain, and genetic engineering are the wave of the future that they have ignored a vastly more important area; basic human psychology and environmental forces.

The illusion that biology is more important than psychology overwhelms the mind of those who inform the public. Few in the media have heard the other side of the story.

Yet, even as the medical profession learns that psychological events in the environment cause many forms of medical problems, pharmaceutical companies and the media have kept up the myth that depression and other problems are caused by our genes, biochemistry, personality type, or personal decisions. The effect of psychology and forces in the environment has rarely been made public.

If we want to understand what makes our lives better, we must first understand what makes us worse; what the causes of human psychological problems are.

JAMA: PSYCHOLOGICAL PROBLEMS ARE CAUSED BY… PSYCHOLOGY?

In January of 2010, a major bombshell was dropped on the world of psychology and psychiatry by JAMA, the prestigious Journal of the American Medical Association.

After a brief flurry of interest, it was ignored by the news media, which seemed to have no clue as to what to do with it.

Psychologists reviewed six major studies comparing major antidepressants to placebos," sugar pills" that have no effect whatsoever. They found that the antidepressants (SSRIs, such as Paxil, Prozac, Effexor, etc.) that have been touted by the media, pharmaceutical companies, and psychiatry and psychology alike for a quarter of a century as the "cure" for depression were only slightly better than a placebo.

News headlines read, *"Studies Show Antidepressants Work Best on Those with The Worst Depression."* No, that is not what the studies said. The studies said that antidepressants do not work at all, or only one point better than a sugar pill, except on the most severely depressed.

People suffering from depression were divided into four categories: "mild," "moderate," "severe," and "very severe"; only those in the fourth category of "very severe" benefited from the medication. Even this group seems to have benefited more from the placebo effect than from the medication. The major placebo effect seems to be psychological; that is, having a doctor who seems to care about us makes us feel hopeful. Hope provides relief from psychological pain.

I have spoken to several psychiatrists in seminars on the subject since then. Some just shake their heads. Some say that the medications do actually work, but their effectiveness is not seen because the effect of the placebo is so much greater that it masks the drug's effect. That is not exactly an overwhelming endorsement of medications.

Others say that you can tell that the medications do work because if their patients go off them, the patient will get much worse. However, this is not a clear test because the body takes weeks to adjust to the medication in the first place; to develop a tolerance for the SSRI medication. When withdrawal is done "cold turkey" without tapering them off, the body does not react well to this. Negative symptoms may occur as the body struggles to readjust, making the patient think they cannot do without the drugs.

If you want to change, talk to your doctor about tapering off as a "drug holiday" trial to see how you do.

The "drug holiday" is the *only* test we have to see if the medication is working. If you get worse, then the drug may be helping. If you get better without the drug, then you should probably never be on the drug. If there is no difference, then the drug may not be helping.

Most of the news media had no idea what to do with the question of the value of antidepressants. It was followed by an almost surreal silence. No one wanted to criticize an entire profession, psychiatry, or an entire industry of pharmaceutical companies, whose livelihood depends on the use of medication, about 80 billion a year, just on antidepressants. And after all, pharmaceutical companies spend big bucks advertising in the media.

Not until almost two years later did Leslie Stahl on *60 Minutes,* CBS's famous news magazine, have the courage to deal with the problem openly. Leslie presented the evidence and then the rebuttal. The psychiatrists who disagreed with the study argued that antidepressants do work, about 14% of the time. Compared to a placebo, which works with depression almost 50% of the time, this is nothing like the miracle cure we have been led to believe by pharmaceutical companies and the media, who presented only glowing positive stories for forty years. Never did we hear it criticized.

So, what is it that causes the psychological problems of living? Why is it that *half of all Americans* will report feeling so bad and distraught that they will "seriously consider" suicide at some point in their lives, even though few will ever attempt it? Before we can control our minds, we must explore what causes the simple problems of living.

Even when we see psychological disorders that are biological in origin, like schizophrenia or ADHD, that is not the whole story. Children with these problems are different. What do other children in public schools do to those who are different? They ostracize, ignore, and bully them; they put them down, and they make fun of them to their face (unlike adults, who do all that behind their backs).

Even the rare schizophrenic who takes a gun and starts shooting people is not doing this because he is "paranoid"; he has grown up in a society, in the school system, where he has been treated like dirt, humiliated for other's entertainment and made to feel like a non-person. Is there any wonder they may become paranoid long before their problems are evident as adults? Is their behavior caused by the biological disorder, or by how society treats them?

We know that children with ADHD have low self-esteem, not because of their biology, but because of years of being told by teachers and parents to "sit down," "be quiet," "stop it!" Ask the adults who went through this as children. They come to see their behavior as their fault. The negative emotions become a part of their brain. They have been changed for the worse by other people's reactions at home and in school.

Even when genetics and biology are the primary cause, they are not the only factor. But in this book, we are dealing with the "ordinary problems" of life and the forces in the environment that affect us all, not with biological problems.

So, if antidepressant medication does not cure a "chemical imbalance" in the brain, then it cannot explain the cause of most of these human problems in the first place. On the other hand, a more powerful psychological force, the placebo effect, just might provide a clue.

THE LINK BETWEEN BIOLOGY AND PSYCHOLOGY

The idea that depression was caused by a "biochemical imbalance" was invented by pharmaceutical companies. It worked amazingly well, with over 80 billion dollars going into sales worldwide every year. Psychiatrists quickly picked up on the idea because it seemed to make sense, and after all, would reputable pharmaceutical companies lie just to sell pills?

We were told that depletion of serotonin, a neurotransmitter, one of many dozens, altered the firing of nerve cells and caused depression. The problem with this logic, despite the lack of empirical evidence, was that even if we did find a depletion of serotonin in people who were depressed, that would not mean anything by itself.

Imagine this: you are in a car speeding along in the rain, and an 18-wheeler pulls out in front of you. You hit your brakes, you skid and skid... Then nothing happens; you do not hit the truck after all. What happens in your body then? Adrenalin and epinephrine shoot into your bloodstream; your heart jumps, and you breathe heavily; your heart rate goes over 100 beats a minute, and your hands get cold and clammy. Why?

Now, suppose doctors do a simple blood test on everybody who has or almost has a traffic accident. And they find overwhelming evidence that such people had 600 to 800 times more adrenalin or epinephrine in their blood. Does that prove that the presence of these chemicals caused people to have traffic accidents?

This is a fear reaction, triggered not by biology but by psychology—the fear of almost dying. But that is a learned fear, triggered by a conditioned association between ideas in your mind. You had no accident—no physical damage to you happened. A learned fear of having an accident triggered the neurotransmitters. A simple psychological association between accidents and the emotion of fear triggered the adrenalin that shot into their blood.

The same is true of every finding that there may be a biochemical cause of a psychological problem. Psychology can control our biology. There is no reason to assume that depletion of serotonin or any other neurotransmitter causes a psychological problem. But that has not prevented the media from being certain that biology is more important than psychology.

THE DIRTY LITTLE SECRET OF PSYCHIATRY AND PSYCHOLOGY:

Inferiority feelings motivate bizarre behavior. Both psychiatry and psychology have always felt like they get no respect from the media. Because they wanted more respect, they desperately tried to be more like the medical profession. They managed to get psychological problems like depression, which was considered a personality trait called melancholia, changed into a "mental illness" in the

1950s— probably one of the greatest mistakes in our history. This transformed a trait into an illness that now carries with it a stigma it did not have in Abe Lincoln's time.

The Dirty Little Secret of psychology and psychiatry is that you cannot get paid unless you give someone a medical diagnosis. If you want to treat a client, you must first label them with a diagnosis, such as Major Depressive Disorder. If you do not stick a label on them, then no insurance company or Medicaid will pay for treatment, except for the initial interview. Just like diabetes or high blood pressure, no doctor can prescribe medication or counseling without first giving people a label.

We have taken the brain's normal response to the problems of life and made it into a negative stigma that may interfere with getting a job or make it impossible to get life insurance. Professionals may argue that it should not be a stigma, but that argument carries no weight in the real world.

Diagnoses are a dime a dozen. The DSM-V has diagnoses for almost any conceivable diagnosis, and many that are quite inconceivable. One such diagnosis is Trichotillomania. That is a compulsion to pull out or even eat one's own hair. I only know of this because I had a client with this diagnosis. I had to look it up. Why do we even need such a diagnosis? So, psychiatrists and psychologists can charge insurance companies and Medicaid for treating the disorder. You can't charge money for treating someone who eats their own hair, but if you make it a mental disorder and label it Trichotillomania....

The more important issue is what causes such a condition? This is from the same phenomena that cause nail-biting, or stuttering, uncontrolled nervous tics, or compulsive handwashing, or eating comfort food, or "shop till you drop". This is a long series of different reactions to anxiety or depression. Why this particular behavior? Who knows? But it is an anxiety reaction. Those who have these traits, display such behavior dramatically more frequently under anxiety or pressure. Lacking a worthwhile way of identifying the causes, the DSM resorts to merely labeling the symptoms.

THE PROPHYLAXIS PROJECT

Prophylaxis is a good classical medical term; it means to prevent disease. Back in the 1950s, when condoms were sold behind the counter at the neighborhood drugstore, it was deemed somehow uncouth or insufficiently genteel to use the word "condom" in public. This led to the cute use of the term "prophylactic" to replace the obscene word "condom"; hardly the first time in our history that culture and emotion have combined to change the meaning of an otherwise good word. That fact, in itself, is a profound comment on human nature and the ease with which words, paired with an emotion, can come to change our very perception of reality.

Television is rife with stories of the latest drugs and medical treatment and the newest medical miracle. Most of this is the usual hype that results when the media is more interested in sensationalism that draws viewers' attention than in medicine's less glamorous reality. Yet, it is a simple fact that preventing medical problems, however unglamorous it may be, is infinitely more effective than treating them once they begin, even with our impressive high tech do-dads.

Modern miracles to treat heart disease may look impressive, but they come with great risks and uncertain promises. Preventing heart disease and stroke by lowering blood pressure, reducing cholesterol intake and environmental stress, and eliminating trans-fats from the environment are more effective, less painful, and much, much easier.

With human problems of living, the media touts psychotherapy as the answer, along with taking a pill. All psychology is devoted to treating clients *after* they have a problem; virtually nothing is done to consider how to prevent those problems. Psychologists do not make money by preventing problems, only by treating them *after* they arise.

Of all I have seen in psychotherapy textbooks, there is little or nothing devoted to using education to prevent psychological problems. Recent articles on preventing psychological problems include one from the World Health Organization, some from William McFarlane, suggesting that we could go into schools and use psychotherapy and medication on "at-risk" students who are targeted for having problems.

But that is not what I am talking about here. The last thing we need is to target specific individuals for pills and therapy; to label them with one more thing that makes them different and a target for name-calling, bullying, and put-downs, leaving them with the unholy stigma of a mental health diagnosis. What we need is to educate all students and adults as to what they need to know to do well in life, what causes the problems of living, and how our own behavior affects others. That is not being done or even considered.

To succeed in the project of preventing psychological problems, we need only a small-scale "Manhattan Project" with the best minds in the psychology field coming together with the ideas necessary to put into educational programs, videos, and internet sources that can be used as teaching tools. Put the knowledge, along with videos and examples, on DVDs and the Internet that can be used by teachers in classrooms across the country. Show them the examples they need to understand how their mind works and let them understand that others have been through these problems before and gone on to succeed. This should not be a single elective class but needs to be repeatedly emphasized over years.

Because foundations for teaching already exist in public schools, this would be incredibly cheap compared to the expense of medicine and therapy for those who need it. We could help our young for less than the cost of a latte a day. Moreover, the value of the program would apply to every individual, not just those with diagnosable problems. Yet, with all the money and attention going to entertainment, nothing is left for helping our children or ourselves. There is no money, no organization to sponsor an educational form of preventive psychology, and little voice even within psychology itself.

WHY HAS PSYCHOLOGY BEEN SUCH A MONUMENTAL FAILURE?
The Future of Psychology as a Preventive Science

Like psychiatrists and counselors, psychologists do not get paid for preventing society's psychological problems; they only get paid for unscrewing people *after* society has screwed them up. There will always be a need for psychotherapy.

However, the future of psychology has to be in preventing problems before they arise, not in treating them after it has happened. This can only occur when we begin to educate children as to the origins of their own minds, to understand where their fears, guilt, and desires come from, the relativity of their value judgments of themselves about others, and give them the skills they need to survive in the real world. That must start in schools. Psychology must become an educational science.

USING THE MEDIA FOR GOOD INSTEAD OF TRIVIA

The news media has enormous power to do good in our country. We saw this firsthand when we watched Martin Luther King and his people in Selma, Alabama, where police turned fire hoses on them, and police dogs attacked them when all they were doing was peacefully marching to protest the lack of freedom in America. The media became outraged at the police's Southern politicians' behavior and quickly began to give enormous coverage to this issue.

The headline coverage went on daily for years. This made it impossible for politicians to stay silent. Until the media gives as much attention to the problems of living and suicide and psychological problems as it gave to the civil rights movement, covering it day after day, month after month, year after year, getting in the face of politicians and asking the hard questions, then nothing will ever be done to prevent this problem.

Not a day should go by without someone in the media sticking a microphone in the face of a politician and asking him or her, in front of millions of people; "What have you done today to help our children?", "What have you done today to help our homeless veterans?", "What have you done today to make this country better?"

"What have you done ...?"

PREVENTIVE PSYCHOLOGY AS A UNIVERSAL NEED

In 2009, the Federal Government's National Academies of Sciences issued a 500-page report on how to prevent mental and behavioral disorders. A major conclusion was that we should intervene early in life to identify and provide treatment to those at risk for mental illness using rigorous screening programs. In other words, single out individuals and get them medication and therapy. This has been the approach of most of what passes for "preventive" therapy.

No, no, no! This is the equivalent of forcing the victim of bullying to get help instead of dealing with the cause of their problem. It means you will further stigmatize the individual who is "different," give them a label from the DSM-V, and ensure that they will feel even more inadequate and worthless. That is nothing like what we need to do. Stigmatizing them by putting them on meds or therapy will only make their problems worse.

Instead of individual psychotherapy, which our system will never be able to afford, we need to use the educational system to provide the education for everyone that will allow him or her to know what to expect and prevent problems from the beginning; to take control of their own minds.

What we need to do instead is to let people understand that most problems of those we have stigmatized as "mentally ill" differ only in degree from the problems we all face in life. Everyone needs to come away with an understanding of the problems that most of us will have to come to grips within our lives. Everyone needs to learn the knowledge, skills, and understanding needed to succeed in life. Psychology must become an educational science, not just a therapy.

In 2010, the World Health Organization published an outline for preventing psychological, emotional, and behavioral problems, mostly by going into schools and identifying those in need, providing them and their families with help.

Is this likely to happen? Would parents put up with singling out children, labeling them, and making them get help? Probably not. Could we afford therapy for everyone who needs it? Probably not. Do we want to label children with a stigma that will follow them for the rest of their lives and make them targets for other people's intolerance? Certainly not.

This is not what we need. We can use the existing educational system to educate everyone about the things that provide some understanding of the problems they are likely to encounter in school or later in life. We can use the students themselves to do good, instead of dumping value judgments on each other. That, too, will encounter resistance. Sadly, it might be another fifty years before a meaningful program for schools will be approved, but there are things parents can do that we will cover later.

SUICIDE STATISTICS ARE ONLY THE TIP OF THE ICEBERG
The Real Issue Is One of Enormous Unhappiness in America

At some point in our lives, the majority of Americans will "seriously consider" suicide. Almost none of those who consider suicide will ever actually attempt it. Few of those who do attempt it will succeed. Yet, this speaks of a profound degree of unhappiness that pervades our culture. And nothing is being done to change it.

THE GOOD NEWS:
If You Fail, You Will Succeed

The good news is that of more than half a million suicide attempts each year that *fail*, only 10% of males and 3% of females will go on to commit suicide. If we get them help, if we let them know that we care, if we can give them goals and hope, treatment does work. Yet, we never get them help until *after* they attempt suicide.

We could easily save half of the 5,000 teens and young adults that commit suicide each year if we cared enough to try. But nothing gets done. The only way to get to those individuals is to start where the problems begin—in the school system itself. But that would be too controversial, so nothing gets done.

THE BAD NEWS:
Nothing We Have Tried Has Worked

The bad news is that in the 1970s, an average of 25,000 Americans committed suicide each year. Now that figure is over 47,000 a year. Some of the increase may be due to population growth. Yet, we now have about six times more psychologists, psychiatrists, and counselors than we had in the 1970s, and also about six times more antidepressant drugs, and we have not made so much as a dent in preventing suicide. To deal with this problem, we would have to start prevention in the public-school system where these problems begin, not wait for them to attempt suicide.

PSYCHOLOGY AS AN EDUCATIONAL SCIENCE,
NOT AS THERAPY

I spent fifteen years working as an associate clinical psychologist with state and community departments of mental health and mental retardation. I have worked with six of the best of the new generation of psychiatrists, and I have great respect for their knowledge and wisdom. The basic rule we saw is the "rule of thirds"; that psychoactive medication works well about a third of the time; it does not work at all a third of the time, and the remaining third it works more or less well. That was not the view that people were fed in the media. If the medication works for you, use it, but it is not a magic pill for most people.

However, no psychiatrist or psychologist in clinical practice ever does placebo-controlled studies. Everyone in practice relies on the studies done by the pharmaceutical companies who push the medications. And there lies the lie.

As far back as the 1970s, British psychologist Hans Eysenck carried out extensive studies of the therapies used by psychologists on patients. He found that, like the anti-depressive medicine used by psychiatrists, psychotherapy fared no better than a placebo.

Yet, both psychology and psychiatry have since been able to overcome the shortcomings of their methods. How? Simply by ignoring the studies. If no one is talking about the problem, if it does not appear on the news, then it cannot possibly be a problem.

Since then, psychologists have started doing more real medicine by comparing their methods to the use of placebo control. Of all the psychotherapies, behavioral and cognitive-behavioral have proven successful, and basic techniques such as counter conditioning and desensitization are at the top of the list, with as much as an 85% success rate with some problems. Not that the others are failures, but most have not bothered to subject their methods to scientific validation.

I have spent twenty years teaching psychology in universities and colleges, and I have seen an enormous number of students who have written their pain and suffering into their papers and spoken openly of them in class. I have seen how little success we have had in dealing with the real problems of life, either with psychotherapy or medication.

I have routinely reviewed new psychology textbooks for major textbook publishing companies, yet I see no change in what we tell students. Cataloging thousands of seemingly unrelated studies that point to nothing and avoid the serious issues of living is a failure of our educational system, a failure of psychology itself. The DSM-5 catalogs lists of symptoms to label problems, we could do better if the DSM listed the many possible experiences that led to patterns of behavior that we have labeled "illness." Yet we are a far cry from doing so.

CHAPTER 8

THE GREAT PSYCHOTHERAPISTS

WHAT WE CAN LEARN FROM THE GREAT PSYCHOTHERAPISTS

BEHAVIORAL AND COGNITIVE BEHAVIORAL:

COVER-JONES, SKINNER, BECK, ELLIS...

Albert Ellis, one of the great cognitive behavioral psychologists, tells a story of how, when he was a skinny, gangly teenager, he had a terrible fear; a phobia. Some people are terrified of snakes, bugs, heights, or enclosed spaces. But Ellis was terrified of women.

Actually, he was terrified of rejection. It is, in fact, the most commonly treated phobia. More people come to see psychologists because of social phobias than any other kind. If you are afraid of bugs, you can call the exterminator. What do you do if you are afraid of people? We saw before that the fear of speaking in public was picked in the Book of Lists as the greatest single fear in America. It has its origins in the fear of being "put down" or being laughed at in the school system. A recent study found that social phobias are twice as debilitating in the workplace as depression. This is perhaps the most effective therapy to use with the greatest problem of our society, social phobias.

Ellis hit upon an ingenious solution; it is one right out of the studies of counter conditioning and gradual desensitization by Mary Cover Jones, a graduate student of John B. Watson, although Ellis came up with this idea on his own as a teenager. First, Ellis decided he was going to start by saying "hello" to just one woman on day one; that's all, just "hi;" something you might say to someone you casually pass in the hall at school. On day two, he would say "hello" to two women. On day three, he said "hello" to three women. By the end of the week, he had said "hi" to about 15 women. Guess what? No one slapped him. No one had him arrested. No one put him down or made fun of him. Most of the women said "hi" right back!

That is a process of gradual desensitization. We do it to ourselves all the time in real life. Have you ever had an accident, or almost had an accident, or even gotten a ticket? That is a scary experience. For the next couple of weeks, we tend to drive reeeeeally carefully. But after a while, we desensitize ourselves by not having an accident or getting another ticket. This is dirt-simple psychology, and it works. We do it all the time without realizing it, but the importance is that the same techniques can apply to any fears.

We have reprogrammed our minds. We come to perceive reality as not so scary, after all. We come to relax. Our fear goes away.

The second week, Ellis decided he would do more than just say "hello;" he was going to ask a question: "What do you think of the teacher? What do you think of the weather? What time do you have? What do you think of the (news)?" He started on day one with one woman; on day two, with two women, and so on. By the end of the week, he had carried on a modest conversation with most of the women.

In the third week, Ellis had a script, a plan of exactly what to say, that he would use on each encounter. And at the end of the third week, he was carrying on a fairly good conversation with most of the women. Not that every encounter was as good; some were better than others, but the fear of rejection gradually went away. Later, when Ellis became a famous therapist, those who met him sometimes noted that he had succeeded so well that you could not shut him up. There is such a thing as overlearning a good thing.

This is one of the dirt-simple tools of psychology that works; it rarely has any negative side effects, requires no long-term use of anti-anxiety medication, and has an exceptional track record of success. Eighty percent of individuals who go through such therapy are cured. It probably would be one hundred percent, but some decide they would rather have the fears than to face them—or perhaps they get stuck with a therapist who is not skilled in basic behavioral therapy. Sometimes we run into people who are just mean to us for no good reason. So, no method is perfect, and it probably would not work so well with bullies, although it might.

But the important point to me is that we can do it ourselves. We do not have to pay a psychologist one hundred fifty dollars an hour to do therapy on us (pardon the professional heresy). We do not have to medicate ourselves with psychopharmacology or alcohol to face our fears. We can learn whatever social or desensitization skills or strategies of living that give us the ability to function well in the real world. Once we understand the underlying principle, we can formulate a plan that works for us.

What Ellis did, not only reduces our fears of rejection, but it is a plan that helps us gain social skills. Then do what actors and actresses do—plan out a script to follow, imagine what you will say and how they might respond; from this, you gradually get positive reactions from others and learn how to keep up a conversation—all skills that are a first step in becoming successful in life. Psychotherapy often uses the same technique of desensitization, but it typically leaves out the more important part, the learning of the social interaction essential to success in life.

Therapists could teach the simple behavioral techniques of Albert Ellis to their clients. It could be of great benefit. Yet far too many are only using the talk therapies they learned in a counseling class.

And the rest of us could teach such simple skills to ourselves or our children.

FUNCTIONAL ANALYSIS OF BEHAVIOR:

In basic behavioral psychology, you start with a Functional Analysis of Behavior. Simply put, you want to know what stimuli are associated with what responses so that you can evaluate what is happening. One example of this came from a psychologist many years ago, whose daughter was afraid of the dark. Now, this is a common fear among children, but parents rarely do a functional analysis of what is going on in even such a simple fear.

He simply looked at the sequence of what was happening. First, he would tuck his daughter into bed, and then second, read her a bedtime story. All of these elicited positive emotions from his daughter. Then, third, he would say goodnight, turn out the light, and leave. Turning out the light (the dark) was associated with an unpleasant, even fearful, event—his leaving. Now the dark came to be *associated* with her security blanket leaving; a very unpleasant event.

So, he decided to simply rearrange the order of the stimuli. Now, he would first turn out the light. Then, he would secondly tell her a bedtime story in the dark. Now, the dark became *associated* with the positive emotions of telling her a bedtime story. She lost her fear of the dark.

Not every issue is so simple. Some children may be afraid of the dark for different reasons—because their older brother told them Freddie Krueger was hiding under the bed and would kill them in the middle of the night, or because they saw a Hollywood film about a ghoul hiding in a closet. There are often no magic bullets. But psychotherapists are still useful simply because of their experience and skill in being able to see the relationship between stimuli and responses that often elude those who are not so trained.

Much of what we encounter in life seem to us to be totally beyond understanding. Only when we think in terms of Stimulus-Stimulus or Stimulus-Response associations do these seemingly incomprehensible behaviors become more apparent.

To learn the skills of a simple Analysis of Behavior is profound. You have seen this happen in the TV show "Supernanny." Although Jo claims she gets her inspiration from "intuition," she very clearly observes the relationship between stimuli and responses in her show. What does the child do (Stimulus)? How do the parents respond (Response)? What does the parent say (Stimulus)? How does the child respond (Response)?

Learning such a simple analysis of behavior is critical to understanding all human behavior in children or adults. Teaching this in our schools would give us a critical bit of knowledge, a tool, to use in childrearing and human understanding. Yet most of us go through life only dimly aware of how our behavior, attitudes, body language, and words affect the way others around us see us. We do not see it until someone points it out.

Learn to use this technique in your daily life. Those who pay attention to how stimuli from others, their words, their subtle expressions, affect your own emotions, and begin to see where their own problems lay, and work to desensitize themselves. Those who pay no attention to this simple Stimulus-Response effect, will often be unable to understand what is happening to them. Understanding that, is essential to getting in control of our own mind.

COGNITIVE BEHAVIORAL THERAPY CBT

Many Cognitive-Behavioral Therapists say that you can treat someone with anxiety or depression or similar problems using reason and comparison. They may ask the individual to make a list of what bothers them the most. Then they may make a list of their bad points and a second list of what their good points are. By reasoning with them it should then be possible to convince them that they can change the way they see themselves, change their understanding of what is good about their life.

Showing people their good points, their successes, their accomplishments is important to changing their self-perception. It can be a valuable adjunct to therapy. But it is not clear how long-term reasoning can succeed in reducing depression and anxiety.

They claim an 85% success rate. Yet no clinician ever does experiments themselves. No doctor conducts controlled studies, they rely on studies done by the pharmaceutical companies. No psychologist does controlled studies themselves; they rely on others. Much of what CBT claims does not come from controlled studies, it comes from having the client do a self-report checklist of how they feel about themselves before therapy begins. Then, after therapy they do another self-report checklist and, guess what, the client feels much better after therapy.

When someone comes in for therapy they tend to be at a low point in their life. After doing CBT for a few weeks, they then take a second self-report and report feeling much better about themselves. I have no doubt their feelings are valid, but how much of that is due to the therapist's positive attitude, to learning to think about their good points, and how much is due to CBT reasoning itself? We need more independent, placebo-controlled studies to figure out what is really happening in these studies, yet there is no money for controlled, double-blind placebo comparison studies.

PSYCHOANALYTIC CONCEPTS: Freud, Jung, Adler...

OUR DEFENSIVE BEHAVIOR INHIBITS OUR ABILITY TO UNDERSTAND The first step in understanding the crippling effect of our broken tail, our inability to function as effectively as we could, is to understand one of the most common of all human behaviors—ego defensive reactions. Only about five percent of academic psychologists today would classify themselves in the "psychoanalytic" camp, along with Freud and others. Yet, of all that Freud wrote, one area stands out as the most important area of all, the area where perhaps his greatest contribution to human understanding was—his analysis of our ego defenses.

Other greats in psychology, such as behavioral psychologists John Dollard and Neal Miller (of Biofeedback fame) rewrote the ego defense mechanisms, interpreting them in behavioral terms instead of psychoanalytical ones. Humanistic psychologists such as Carl Rogers and Abe Maslow wrote them into their psychotherapy as one of the stages through which an individual must pass on the way to becoming a better person, the highest level of self-actualization. They are an essential first step in human understanding.

"Life as we find it is too hard for us; it brings us too many pains, disappointments and impossible tasks. In order to bear it we cannot dispense with palliative measures...

"There are perhaps three such measures; powerful deflections (e.g., ego defenses) which cause us to make light of our misery; substitutive satisfactions (e.g., entertainment) which diminish it; and intoxicating substances, which make us insensible to it (e.g., drink and drugs)." (parentheses added) Sigmund Freud

THE EGO DEFENSE MECHANISMS

RATIONALIZATION:
"I made the best of all possible choices."
No matter what we do, we can create a "rational" reason as to why what we did was the best possible choice. "I can't see anything else I could have done" is a common statement. "It is all for the best," is another.

Suppose you have a choice between buying a new car or keeping the old clunker a few more years. If you buy the new car, you tend to rationalize, "Yes, there is a heavy monthly payment, but it won't break down, use up oil or wear tires thin, and it may get better gas mileage. And, it is a great 'chick magnet.' So, I made the best possible choice."

On the other hand, if we decide *not* to buy the new car but to keep the old clunker, what do we do? We rationalize that, "Sure, it is old, but at least I won't be stuck with that heavy monthly payment; it won't depreciate thousands of dollars as soon as I drive it off of the lot, and it doesn't hurt as much if I get a scratch or dent in it. So, I made the best possible choice."

No matter what we do, what we believe, or who we vote for, we can convince ourselves that we made the best possible choice. At a more personal level, suppose we apply for a promotion at work. If we lose out to another employee, we may rationalize, "He kissed up to the boss," "the boss showed favoritism," or "my good points were overlooked."

We may also rationalize that, "It is better I did not get the job because it takes a whole lot more work, has more stress, and I don't like the new boss anyway." The latter is known as "sour grapes" rationalization after the Greek fable about the fox and the grapes. The fox was wandering through the forest when he spied some luscious grapes hanging from a vine. He jumped and he jumped, but he could not reach the grapes. As he dejectedly walked off into the forest, he said to himself, "Oh well, they were probably sour anyway." So, he made himself feel better (avoided tension or anxiety over failure) by thinking the grapes were sour.

Suppose two guys are trying to win the same girl. One gets her; the other does not. As the failed suitor dejectedly walks off, he thinks to himself, "Oh, well, she would have been a bitch anyway."

There is a rationalization for every occasion.

Of course, sometimes our rationalizations are right. Sometimes the boss did show favoritism, or the job is more stressful than the one we have. But we make such rationalizations far more often than the evidence suggests is ever correct.

DENIAL OF REALITY:
It Really Didn't Bother Me Anyway.

In a "Peanuts" cartoon, Charlie Brown is shown telling Linus, "I have found that no problem is so big or bad that it can't be run away from." That is the essence of Denial. Denying there is a problem is like running away from a problem, in our mind.

One of the most remarkable examples of this comes from when Speaker of the House Tom "the Hammer" DeLay had to step down from his job because Ronnie Earle, a district attorney in Texas, indicted him. When his mug shot was to be taken at his arrest, he showed up wearing a suit and tie and a broad smile; for a mug shot? He knew the picture would be presented all over the news, and it made him look like he was posing for a portrait rather than a mug shot. When CNN interviewed him, he appeared happy and jovial.

Newswoman McClusky repeatedly asked Tom DeLay if it bothered him that he lost his post as Speaker of the House. He repeatedly denied that it did. Again and again, she kept asking him if it didn't bother him. Again and again, he kept denying it bothered him at all. Yet, he put out an ad attacking the district attorney who prosecuted him, comparing him to a mad-dog prosecutor.

"Denial is not a river in Egypt." Mark Twain

Anna Freud, Sigmund Freud's daughter and also a psychoanalyst, wrote the book on Ego Defense Mechanisms. Of denial, she says that it is amazing how often we make use of this mechanism in our interaction with children. Contrary to obvious fact, we often tell them that a bit of food they do not want to eat is "really quite good." Or that the immunization shot they just got, that hurt, wasn't bad at all."

DISPLACEMENT:

Displacement of aggression is when we are angry at someone but can't express our anger toward them; as when a boss berates us at work. We cannot tell the boss what we think of them, or we might get fired. So, we take the anger home with us, and when the wife or child says something innocent, we may explode, not because we are angry at them, but the anger comes out on someone who can't fire us.

The best example of this I ever saw comes from anthropologist Sherwood Washburn. He was studying the dominance hierarchy in chacma baboons in Africa. To observe the hierarchy, they were tossing bits of meat to the baboons to see who got the meat. The dominant Alpha males would always get the meat. Behind them were a group of lower-ranking males who also attempted to get some of the meat. But if a lower-ranking male started toward a piece of meat thrown in his direction, the Alpha male would also start toward it, and the lower-ranking male would back off.

During filming, one of the lower-ranking males would show his frustration by repeatedly slamming the ground because they were afraid to slam the Alpha males, who might beat them up. After several failed attempts to get the meat, the frustrated male suddenly attacked a peaceful female who had been munching leaves nearby and chased her up a tree. Now, the peaceful female had nothing at all to do with the problem, but he was angry at the alpha males, who he did not dare attack, so he took his anger out on an easier target.

The human variation of this is a man, who is berated by his boss at work but does not dare tell his boss what he thinks of him because he might get fired. So, he takes his anger home, gets a beer out of the fridge, and when his kids or wife says something innocuous, he suddenly gets mad at them.

PROJECTION: Projecting our Own Motives and Behavior to Others

Projection, in its most common meaning, is considered projecting the blame for our own failures on to others.

Or," He hit me first!" Overused excuse for getting into fights.

Or, "I would have won the election, except..." Overused excuse for losing.

Or, "It's not my fault hundreds of thousands of Americans are dead, it's the China virus." blame China.

Or, Whistleblowers may be accused of being "disgruntled" employees, as if that discredits their complaint.

But Freud's original use of the term applied to projecting our motives onto others.

People and politicians often see their own unacceptable motives in the behavior of others. Those who became "whistleblowers" to expose problems in the office or in Washington are often accused of being "politically motivated" or lying, when the political motivation and lies of the accused are being projected onto the whistleblower.

People who are pathological liars, tend to believe that everyone else is also a liar, thus excusing their own lies in their own mind.

People who have no empathy or feelings for the people they bully or otherwise hurt tend to assume that everyone else is the same way they are and others are just faking caring about others, because that is what they do.

HUMANISM: Rogers, Maslow, May

Humanism in psychology is not quite the same as Humanism in philosophy. Carl Rogers named this area Humanism to emphasize that the focus of psychotherapy should be on humans, not on rats and pigeons, as Skinner and Watson had. Although it must be noted that we learned a great deal about humans from studying rats and pigeons, the basic ideas of desensitization, counter-conditioning, and retraining the brain came from behaviorists.

Of all the psychotherapies, and there are many, Humanism has probably had a more dramatic effect on contemporary therapy than any other, certainly more than psychoanalysis.

First, Humanism is "non-directive" therapy. The client should determine the direction of the therapy, not the theory of the psychoanalyst. Rogers, and also the behaviorists, argue that Freudian psychoanalysis is directive therapy. Freud was interested in dreams and the role of sexual energy in producing problems. So, if a client mentioned something about dreams or sex, Freud would lean a little closer, ask more questions. If they talked about other problems the Freudians might lean back and yawn. The client learned that if they talk more about dreams and sex, the therapist would pay more attention to them.

Alfred Adler, also a psychoanalytic theorist, was most interested in the Will to Power and the conflict, birth order and roles within the family. So, if a client mentioned something about the power roles in the family, he might lean a little closer, ask more questions. That reinforced the client talking about family issues.

Carl Jung, again a psychoanalytic theorist, was more interested in personality types and religious symbolism. So, if a client talked about personality issues or religious symbolism...

Second; Humanism is "Client-Centered" Rogers noted that we are not medical doctors and we should not call the people who come in to see us "patients". This was a revolution in psychology, and to this day the entire field tends to recognize that the role of a psychologist is to assist a "client" in improving their life, not to "cure" them with a magic potion.

Third; Rogers says a basis of psychotherapy is to envelop the client in an **atmosphere of positive regard**. That does not mean that they agree with everything a client does or says, but that therapists must be non-judgmental about the client and accepting of the client as a person of worth. This seems like such an obvious need for a successful client-therapist relationship.

Forth; Abe Maslow, one of the most famous proponents of the Humanistic approach is best known for his concept of a **Hierarchy of Needs**. We have basic needs for food and water at the base. Almost all of us have these needs met. But in extreme circumstances, when these needs are not met, people have engaged in bizarre behavior they would never otherwise even think of.

The Donner Party in the 1800s was stranded in what is now Donner Pass at the beginning of winter while trying to make it to California. The snows were too deep to move. The people ate their oxen, their horses, their dogs, and finally, when there was nothing left, the survivors ate their dead. The same thing happened to a soccer team whose plane crashed in the Andes mountains in the 1980s. Without food, they lived by eating their dead. Hollywood made a movie out of this called *Alive*.

Now that is certainly something none of the rest of us would ever even think of when we live in a land of plenty. But people deprived of the most basic needs will do things they would never imagine.

The important point of Maslow's Hierarchy is that once those lower-level needs are fulfilled, we would *automatically* move into a different level, a higher level on the Hierarchy.

Further up the hierarchy is the need for self-esteem. You see little boys, and politicians, brag about how great they are, how they are better than others, how successful they have been. But that is something we tend to discourage in adults. Until this need to feel respected, to increase our self-esteem, is met, we will engage in ego-defensive mechanisms like Projection, Rationalization, Denial, and more to guard our self-esteem.

This is a remarkable melding of the Ego Defense Mechanisms of Freud with the behaviorism of Dollard and Miller and also the concept of the Hierarchy of Needs of Maslow. Only a few other concepts have been picked up by so many others, in so many different ideas.

The most important point, however, is that once our Ego Defensive Needs and self-esteem needs are satisfied, we naturally move into a higher level on the Hierarchy of Needs.

Finally, after we reach the highest level, the level of Self-Actualization needs. Here, we presumably, no longer need to worry about our self-esteem, about impressing others, about what our boss thinks, and we begin to think in terms of "reaching our greatest potential". Here, humanists talk about "actualizing our greatest potential" or going as high on the hierarchy as we can in our life.

To Maslow, this meant anything from Albert Sweitzer, a doctor who abandoned his practice to go to live in the outback in Africa to bring medicine to those who had no medicine, to Albert Einstein, who abandoned "...*the nothingness of the hopes and strivings that chase most men throughout their lives...*" and instead spent his life trying to understand the most basic principles of the universe.

It would be difficult to come up with hard scientific evidence for the existence of Self-Actualization, but it has a nice intuitive feel to it.

In therapy, Humanism tries to help the individual "achieve his greatest potential". The underlying mechanism is the idea that once our lower needs, such as for respect, for self-esteem, are satisfied, the client will automatically move on to a higher level.

Fifth: REFLECTING THE CLIENT'S FEELINGS BACK TO THEM"

To help the client achieve a higher level of their greatest potential, and also to be "non-directive" one of the techniques championed by Carl Rogers is one of reflecting the client's feeling back to them.

For example: if a client says, "My mother never really liked me. She liked my sister better. She did everything for my sister."

Rogers might say; "So you think your mother did not do anything for you?"

The client might say, "Well, she did help me with my braces, but..."

When I first heard this I thought, huh! they could replace psychotherapists with a trained parrot. But it actually can be a very useful technique. For one thing, it is non-threatening. You are not being judgmental; you are not challenging the client's feelings. More importantly, you are forcing the client to reflect on their feelings, to reevaluate their belief system, to consider what they need to change in how they see their life, to, hopefully, move to a higher level in the Hierarchy of Needs.

GESTALT THERAPY: FRITZ PERLS

The famous saying in Gestalt Theory is that "The Whole is Greater than the Sum of its Parts". Gestalt emphasizes the individual taking responsibility for their own therapy and being aware of the present, not focusing on the past. Freudians and behaviorists emphasized the effect of early experience on the individual, whereas Perls emphasized the present, and your responsibility to take action. Perls was not interested in your past, only in the present.

But the more famous part of Fritz Perl's therapy is that, in contrast to Rogers Non-Directive therapy, Perls is often very directive.

If Rogers would have a client complaining that her mother never liked her, Rogers might say, "So, you think your mother never did anything for you?

In contrast, to the same woman, Fritz Perls might say, "So what if your mommy never liked you. What I want to know is what are you going to do to turn your life around."

There is a famous old black and white film showing the same woman getting therapy from Carl Rogers and from Fritz Perls. At the end of the therapy, the woman said she liked that Carl Rogers but she couldn't stand that Fritz Perls.

Looking back on that film I see now that the therapists, who knew they were being filmed to demonstrate their unique styles of therapy, were trying to represent what was unique about their point of view. It did not necessarily represent what they would actually do in therapy.

But there is an important point here, that there are times when we all need to be told, in no uncertain terms, that what we are doing is not working. We have made ourselves miserable with worry, we have failed in relationships, we are lost in our anger over not being treated fairly. What good has that done? What do we need to do to change?

Fritz Perls put the onus back on us. That is not, of course, the way we would want to start a therapy session.

REALITY THERAPY: *William Glasser*

Reality therapy comes with some similarities to Gestalt therapy, but it has an emphasis on simply confronting the individual with the reality of their situation and inviting them to choose to change.

It is often used in prisons or in detention centers to confront individuals with their choices. It might go something like this; "Look at where you are. You are in a detention center. Within five years of getting out of prison half of you will be back in prison. It is a revolving door. What can you do to change your future? What will you do to keep from coming back here?"

"What can you do to make your life better five years from now?"

Then they would have to follow up with giving them some ideas of how to achieve this: Maybe getting their GED high school degree, or taking college courses, or learning a skill like welding or plumbing.

The problem with the solutions is that once they get out of prison, if they have an arrest on their record, no one will give them a job. Therapy has only limited use in a society that does not care to help. So half of them end up back in prison within five years.

But, as with Gestalt therapy, there are times in therapy when this could be very useful to bring us down to earth with reality.

EXISTENTIALISM: Shakespeare, Soren Kierkegaard, Fredrich Nietzsche, Viktor Frankel

One of the central tenants of existential theory is the idea that man is the only animal that is consciously aware of the inevitability of their own death. This sets up a sense of the need for meaning in one's life and a fear of the meaningless of one's life.

A central crisis in our lives is a confrontation between our feeling of self and our sense of meaningless. This is often called The Existential Crisis. Nowhere is this more evident than in the writings of William Shakespeare from Hamlet, Act V Scene V:

Tomorrow... and Tomorrow... and Tomorrow
Creeps in this petty pace from day to day To
the last syllable of recorded time. And all
our yesterdays have lighted fools the way to
dusty death.
Out...Out brief candle.
Life is but a walking shadow
A poor player who struts and frets
his hour upon the stage And then
is heard no more.
Life is a tale, told by a fool,
Full of sound and fury,
Signifying nothing.

This is an example of man's confrontation with the meaninglessness of his own life. What is it that gives meaning to life? Many things can. Some find meaning in religion. Some find meaning in their work. Some find meaning in their children's future, or in working to support their family, or in helping others...

One way to help someone who is stuck in an Existential Crisis is to help them find something meaningful, short-term goals, and long-term goals.

Most students are preoccupied with short term goals: having a date for the weekend, getting past the next exam. But most do well on their long-term goal, which is to survive the incredibly boring classes long enough to make it through to a job on the outside to an equally boring life but one that pays you money to be bored. Money helps.

A great many people work for a living, and may enjoy gossiping with their fellow workers, but the real goal they have in life, is to work for their family and their kids, that makes it easy to endure life in a meaningless job. Their family is their important goal in life.

Existentialists are absolutely right about the need for goals in life to help give meaning to life.

In one of the most painful experiences in our lives, the death of a close loved one, the most effective therapy is not to take time off of work, but to go back to work as soon as practical. It is keeping busy, getting our brain focused on tasks at hand, that keeps us from dwelling on our loss. The worst time after such a crisis is at night, alone when the thoughts come back, the feeling of loss is greatest.

Viktor Frankl is generally considered a Humanistic psychotherapist, although he created his own therapy known as Logotherapy. But his book, *Man's Search for Meaning,* puts him in the Existential Crisis category. Frankl spent three years imprisoned in a Nazi concentration camp during WWII. He saw life at its most extreme. His brother died and his mother was killed at one camp. He saw that people under extreme conditions of hunger and cold would do things they would never do in normal life. Some would steal food from the weaker. Some became "Capos" prisoners who would keep order in the prison for the benefit of the guards, for a small extra portion of food. His comment in the foreword to his book is the very essence of humanity at its worst.

"Those of us who were there know, the best of us did not survive." Viktor Frankl

In part, Frankl says he survived the concentration camp because of a small *goal*, a glimmer of hope; that if he did survive, he could write about human suffering at its worse from the standpoint of a professional psychoanalyst and neurologist. We need to believe that we can survive, we can get through the worst, to survive.

When I was taken to the concentration camp of Auschwitz, a manuscript of mine ready for publication was confiscated. Certainly, my deep desire to write this manuscript anew helped me to survive the rigors of the camps I was in."

The beauty of Frankl's work is that it shows the reader that, even in the worst of circumstances, there is hope. It counters the emotional feeling that our lives are so meaningless, or that our problems are so great, that we will not survive. Finding a goal in life gives life meaning. For some, religion provides that function. For others it may be helping other people. For Einstein, it was the realization that he could come to understand how the universe works.

No one can tell you what goals will work for you. Probably we never know ourselves until we stumble into a goal. The most important thing is that it be achievable and that there are intermediate steps, mini goals, to get there. But a goal is most useful as an ongoing process. People sometimes find that achieving their goal provides momentary satisfaction, a great high, but after that, what do we do then. There is often a letdown, a need to find a new goal to give meaning to life. You may see this in Olympic athletes who have achieved their dream, been at the top of the pyramid, but what do they do now? Or in mothers with the "empty nest syndrome" where, when their children are married and gone, what do I do now?

Find goals for your life. For students, there are short term goals, surviving another hard class, making it through another exam, all of which leads to their long term goal of graduating into a job that will help themselves and their family do well in the real world.

THE ECLECTIC PSYCHOTHERAPIST

Most psychologists today do not consider themselves to be Freudian or Behavioral or Rogerian or Existential. Some do. But most pick and choose whatever seems most likely to work with whatever problem their client has. Although many think their therapy is likely to work with almost any problem.

But if you had a client who had a social phobia, you might be better going to a behaviorist or a cognitive behaviorist/rational-emotive therapy like Albert Ellis. On the other hand, if you had a client with a feeling of meaninglessness caused by the death of a loved one, an existential therapy that helped you find goals and meaning in your life, might be better. Or go back to work and concentrate your attention on the work you do.

If you have a problem with self-esteem you might do better with a Carl Rogers, Abe Maslow type of therapy. But if you were obsessive-compulsive and making everyone in your family or your employees miserable, you might need a Gestalt or Reality therapist for a "come to Jesus" moment.

Carl Rogers's therapy is a good, non-threatening way to start a therapy session. Other methods may be called in as needed.

Whatever the choice, each of these therapies offer something important to consider.

One should not overlook the major therapy practiced for most of our life on this planet; work. Our ancestors worked hard from dawn to dusk, just to put food on the table and survive another day. Today, we have so much free time, our mind can wander from one problem to another. If we are focusing our attention on something important, that removes much of the time that we would spend worrying. Simply by focusing our attention on something, such as work, or talking to others, can prevent some of the problems caused by dwelling on the problem.

GROUP THERAPY: A Powerful Choice

There are many more choices in therapy, including group therapy where others with similar problems tell their story. That can be beneficial just to find that others can emphasize with you, commiserate with you, understand you.

And there is a wealth of self-help books available everywhere. Even if you have a good therapist, it will often help to find some good self-help books to add to the therapy. Knowledge is essential as one pillar of support.

One of the most effective methods of group therapy is surprisingly rarely used. The videos from Youtube and Ted Talks mentioned in Chapter I are powerful ways of generating discussion and relieving problems within a group setting. Pick one and use it at the beginning of the group setting and it is likely to propel the discussion into a higher level than you could ever get by asking individual members to just tell their story if they feel like it. For many common problems, this should be a first choice.

Group therapy can be a great help. If everyone in the group has the same problems, they may be more understanding of each other than the public. If they learn to support each other, to give each other sincere reinforcement, then the problems can be helped.

PROPHYLAXIS:

PREVENTING PROBLEMS BEFORE THEY BEGIN

Of course, we could do one better, and use the school system to educate our youth to be the first responders, to understand what is happening to them before it harms their self-esteem, to show them example after example of how others made it through their problems, to educate them to understand how to help each other. That is what this book is about.

CHAPTER 9

SOURCE OF THE PLAGUE

Half the time, if just one student said 'stop it,' it would stop the bullying within ten seconds." Psychologist Wendy Craig

The problem is that almost no students will ever say "stop it" or even run to tell a teacher. The fear embedded in their mind by the peer group, the fear that controls their behavior, is being labeled a "tattletale."

SCHOOLS ARE THE BEGINNING OF MUCH OF HUMAN PROBLEMS

Although bullying has become an issue we hear about more recently, it is only part of the school system's problems. We learn to fear what others think, to change our behavior to be liked by others, and to behave in ways that are inconsistent with our own welfare. And we pass judgment on others and ourselves.

Far more problems result from the school system than bullying, yet there is a strange connection between the causes of bullying and that of many human problems. All involve the judgmental attitude so basic to human nature, or at least to our society. So let us explore bullying and what to do about it first.

In Texas, every school system is required to have an anti-bullying program. I guarantee that if you ask any principal of any school at any level, "Do you have an anti-bullying program at your school?" they will all say the same thing, "Oh yes, we have zero tolerance of bullying at our school!"

"What is your anti-bullying program?" you ask them.

"We tell them, don't bully!" they say.

THE KEY TO STOPPING ABUSE IN SCHOOLS:
Wendy Craig's Genius

Telling students not to bully is no anti-bullying program. But, they will insist, they have penalties for bullies; they "get tough" on bullies.

This kind of program is useless as an anti-bullying program for one simple reason—studies show that teachers intervene in bullying less than 5% of the time. Teachers never see it the overwhelming majority of the time. The bullies are smart enough to never bully in front of a teacher; it happens on the playground, at recess, after lunch, on the bus, and before and after class, when there are never any teachers around to intervene.

And the students will not tattle.

So how can getting tough on bullies possibly work when the teachers never see it, and the students won't tell? It cannot.

What does work...

The key to stopping bullying was provided in a remarkable series of studies by psychologist Wendy Craig, who found that:

"Half of the time, if just one kid says 'stop it,' it will stop the bullying within ten seconds."

Of course, you do not want just your kid to be the only one to say "stop it" to a bully—he might be bullied next. The trick is to get enough students on the side of the kid being bullied, not just one. If 20% of the other kids would join in on the side of the kid being bullied, if more than one kid says "stop it" or goes to tell the teachers, it can have a profound effect. No other method is likely to be successful.

THE SILENCE OF THE LAMBS

The problem is that other kids routinely ignore bullying and name-calling— they either laugh at it or avoid doing anything. In studies, Wendy Craig videotaped middle school kids on a playground. In a series of sad examples, she saw one boy being tormented by other kids; otherwise nice kids, who shoved him, knocked him to the ground, and stepped on and kicked him while he was down. All the while this was going on, a group of kids stood only feet away talking to each other. No one said, "Stop it." No one called a teacher.

In another example, a girl was taped who was being bullied by other girls. One grabbed her by the neck, forced her to kiss a tree, and dragged her around the playground. Another girl said to the others, *"We're hurting her, wanna help?"* Another girl told the bullied girl later, "I think she wants to kill you. I think you are going to die." The bullied girl just stood there. No one helped her. No one told a teacher.

Why does no one help? Psychologists who work with middle school students tell us that the worst thing you can be to a pre-teen is a "tattletale." They think it is worse than being a child molester. How can this be? Early in school, they are hit with this: "He got him in trouble. He's a *tattletale.*" And they hear this with emotion. This emotional conditioning of the peer group, by the peer group, takes precedence over any kind of feelings they might have. It rules their minds.

Nor does this end with school; do you remember the Rodney King beating that led to the LA riots? King led police on a chase through the city. When they finally pulled him over, the police dragged him out of the car and beat him over and over again. While lying prone and unresisting on the ground, one police officer stomped on his head. He tried to get up. Another officer used this as an excuse to continue beating him while he was on the ground. The beating went on for fifteen minutes. He had a fractured supraorbital ridge, broken ribs, and multiple injuries.

Of the sixteen police officers involved in the chase, only four actually beat Rodney King; the others just stood and watched. Not one tried to stop them. Not one spoke out or turned them in. And when they wrote their reports, they lied to protect the guilty officers; no one wanted to be a tattletale. Instead, they want to be seen as heroic by the other police because they stood up for their buddies against someone who had been labeled a criminal. It was "Us guys" against "them guys."

If it had not been for a man across the street who had just gotten a video camera for his birthday, no one would have believed Rodney King's version of what happened. They got it all on videotape.

Even medical doctors, presumably well-educated, are reluctant to testify against fellow doctors they know to be incompetent or even to publicly criticize other doctors for medical procedures or drug treatments of dubious value; one reason no one ever publicly questions the use of anti-depressive medication. Childhood experience has bought adults' silence. No one wants to be a tattletale. An enormous amount of bad occurs because we are afraid to speak out; because we know we will be attacked if we do. We know others will blame us:

"He started it!" Conditioned emotional cliché from a guilty ten-year-old?
"Snitches get stitches." Conditioned cliché from a mob boss?
"The FBI should be investigating Hillary Clinton (instead of me)," or James Comey or Michael Cohen's father-in-law, and so on. Conditioned emotional clichés from Donald Trump.

The kids who are bullied, called names, or made fun of by others get no support from other students. They come to feel what Martin Seligman called, "Learned Helplessness." If no one else sees anything wrong, if no one else stands up for them, then maybe they deserve what is happening to them. We see the same thing in child and spousal abuse. They may think the put-downs and taunts are because there is something terribly wrong with them. They learn from the indifference of others; that there is no hope, nothing they can do, no one who will help.

The media always responds with the same cliché—" Get them help!" —but it is a myth that help is easily available or even useful if it is available. The cliché only helps the media feel better about being unable to help.

TREATING THE SOURCE OF THE PLAGUE:

I was amazed to read John Holt's book *How Children Fail,* in which he describes the fears many parents have that school will "brutalize their children." He aptly describes how teachers do just that, often without being aware of it, through their demands for performance and forcing their expectations on their students.

But the brutalization occurs in far more ways than that. The worst form of all is from that imposed by the children themselves. Parents often feel blamed by psychologists for problems that have their origin in the words of classmates and friends. The question is not who to blame, but what do we do about it?

Public schools did not even exist until about 200 years ago. Founding fathers Thomas Jefferson and James Hamilton realized early on that a democracy, dependent on the people's vote, also depended on an educated public, and they were not talking about reading, writing, and arithmetic; they meant understanding human nature, politics, and real history. Until then, only the rich kids could get an education.

But the school system was not put together by educated men who thought deeply of the problems it would create; it simply evolved, gradually and imperceptibly, into what it is today—a vast pool into which we dump our children as a holding tank while the parents go off to do other things, and let them sink or swim.

Instead of teaching the most important knowledge, it dumbs down what students are allowed to learn by pandering to what some have called "The lowest common denominator," with routine censorship of the very information students need to make informed decisions.

This censorship is based on the "politically correct" notion that we cannot teach reality if it disagrees with someone's beliefs. We never call it censorship, of course—we call it "being sensitive to other people's beliefs," "avoiding controversial ideas," or "following school policy." But it is censorship regardless of the words we use to describe it. And it is censorship of the very information students need to function in the real world.

THE MEDIA AS THE BEST PREVENTIVE PSYCHOLOGY

Even though the news media has done a horrendously poor job of dealing with serious issues in suicide and other problems, they have done an outstanding job in some areas. ABC Family (abcfamily.com) is one such remarkable resource, better than anything psychology has produced. You can find serious work such as the *20/20* specials and more on their website; you can even watch them with your kids to help give them insight into what to expect in life as they go through puberty. There is a series about the angst of puberty, *The Wonder Years,* or the problems of bullying with *The In*

Crowd and Social Cruelty. Or the problems parents and teens have in dealing with each other in 20/20's *"Teens: What Makes Them Tick."*

Many have developed programs to stop bullying—from role-playing to using teachers to police the grounds during recess and lunch, watching for bullies. But the best anti-bullying message comes from a John Stossel film for 20/20 called *"The In Crowd and Social Cruelty."* In this film, Stossel and his ABC film crew do an astonishingly good job of showing the causes and potential treatment for bullying— not in the role-playing or other methods he discusses to stop bullying, but in the power of the film itself.

Stossel shows in detail the kids who are being bullied and shunned, telling stories of how they were shoved into lockers, kicked, threatened, and treated like dirt by other students. Showing this film to students beginning in middle school can have a powerful effect on the other students. Students who ignore this behavior when it is going on around them are powerfully affected when they see it happening in this film. They see the pain on the victims' faces, and they hear their anguish and tremor in their voices as they tell their story.

Finally, the mother of a young girl reads aloud the note left by her daughter after her suicide attempt, tears streaming down her face. Her daughter shares how she has always felt an outcast, and how she cannot go on feeling so badly. The school did nothing to help. Her suicide attempt failed. Her mother got her counseling for her problem, but nothing stopped the bullying at school. Counseling did not help. Months later, she killed herself with a gun.

Telling the victims to "get help" does no good. If the school system is unable to change the climate of the students themselves, psychotherapy cannot be effective if the child is sent back to the same environment. Change the environment, not the victim.

Yet simply showing this film, with the power of the mother sharing the story of her daughter's suicide, with other girls telling, with obvious pain, how they were bullied, is so compelling that the film itself can get across to students on the side of the bullies never see. It can bring out compassion in students who then want to know, "What can I do to help?" Without that, there is no way to change the problem of bullying.

Until every student is on the same page, willing to speak out, be a tattletale, call a teacher, say, "Stop it," nothing the schools do to "get tough" on bullies will have more effect than raindrops on the ocean. But this film will never be shown to students in public schools because it includes suicide. Schools are terrified of showing anything that even remotely might be considered "controversial." They would rather let 5,000 teens and young adults commit suicide each year and tens of thousands more turn to drugs, alcohol, or major depression than to risk the slightest hint of controversy; or to have to spend money to help. And it is so much easier to blame the victims, the kids who use drugs or commit suicide, than to accept their own failure as adults to deal with the real problems.

The reason schools are so frightened of even mentioning suicide is that they are afraid that if they talk about it, it will encourage some students to go ahead and do it. That is a serious concern. Yet, most studies show that students who feel suicidal want to hear someone talk about it. They have no idea that anyone else has ever felt that way and survived. We have left them to feel that they are alone, that no one in history has ever felt as bad as they do, that they are the ones who are sick, labeled with the voodoo curse of mental illness. But if even one child committed suicide following such a presentation, the sky would fall; parents would cry havoc. The media would have a field day interviewing the grieving parents. Politicians would demand censorship. No one would care that it helped others, that it saved many.

Of course, it would be easy to take out the suicide from the film, and you would still have a powerful film that makes it difficult for students to ignore the effects of bullying, although it would not be as effective.

THE ROLE OF CONDITIONED EMOTIONS
IN ANOREXIA AND BULIMIA

In the film, they also talk with students who were bullies in school. Amazingly, the bullies talk with some pride about the kids they bullied. One even brags, "I made some girl bulimic." He demonstrates how he verbally harassed her, with a tone of contempt in his voice, by calling her fat and "disgusting" every day and asking how much she weighed—until she started throwing up, trying to lose weight.

Those in psychology who try to find a "biochemical" or genetic cause of bulimia or anorexia should take note; this is a far more common cause of these problems. Psychology has named this, to make it into a disorder—we call it "bulimia," just as if having named it actually did any good.

One of the remarkable points made by Stossel's interview was the dramatic difference between the kids who were bullies, who laughed and joked about what they did, compared to the kids who had been bullied. The kids who had been bullied, even years later, still told their stories with such obvious pain and anguish. This you would have to see to understand.

Yet, even those kids who are bullies do not come away from their experiences in the school system unscathed. Of those kids who are bullies early in school, over 25% will have a felony arrest on their record within five years of getting out of school. In the film, you see bullies shoving kids, stealing toys, pushing a kid to the ground, kicking him repeatedly while still on the ground, forcing a girl to kiss a tree, and inviting others to "hurt her"; they do.

What would happen in an adult workplace if employees shoved, hit, or kicked another employee? They would be arrested and charged with felony assault.

Adults would not put up with this. Yet we see nothing wrong with it when it happens to our kids?

The bullies learn in school that they can get away with bullying other kids. Even adults blow it off as, "Kids will be kids." They try the same thing in the workplace or the real world, and they end up in jail. Their rate of spousal and child abuse and divorce are far higher than average; all because they learned in our schools that they could get away with treating others like dirt, and other students would laugh.

THE NATURE OF BULLYING HAS CHANGED

The nature of bullying has changed. In my generation, if you got to know a kid who was a bully, they turned out to be pretty sad. Typically, they were bullied by their dad or older brother at home and imitated their behavior at school on kids they knew could not fight back. When I was in school, the only kid I knew who was a bully once invited me to his house. While there, his father put him down and belittled him unmercifully. He never dared to talk back to his dad, but when he went to school, he imitated his dad's behavior with other kids he knew he could bully.

Today, they say the kids who bully are the "popular ones" who get away with it because everyone wants to be their friend. I cannot even remember seeing a popular kid bully another in my generation. Those I knew were always pretty nice. What has changed?

In a previous generation, comedians like Jack Benny, Bob Hope, George Burns, and others made fun of themselves or the system. Today, comedians make fun of other people. Judge Judy is famous for her arrogance and mean-spirited put-downs of others. Donald Trump puts anyone down who disagreed with him. Their audience laughs.

Students see that and try it out. They make fun of other students and put them down. And the students around them laugh. The laughter of their fellow students ensures that they will continue to put others down.

Even the age of bullying seems to have changed. I cannot remember ever seeing a kid bully anyone until middle school, about eighth grade. Recently, a student told me about how her daughter, in third grade, came home crying one day. Her daughter said that the other kids were shunning her. They ignored her if she talked to them; they acted as if she did not exist, and they turned away if she tried to talk to them. She had a talk with her daughter about this. The next day, her daughter came home happy. She asked her daughter how things went in school. Her daughter said, "Oh, everything's fine today. Today we are all shunning Susan."

Even the advice we used to give to kids on how to deal with bullying no longer works. In my day, fathers used to tell their sons, "Hit him in the face. That will stop it." The only time I tried it, it worked. However, you cannot tell kids that anymore. Bullies tend to go around in gangs or with a group of friends. They know to pick on the smaller kids, who are alone, who won't fight back, and if one did fight back, they would have the bully's friends to deal with.

WHAT CAUSES BULLYING?

Wolfgang Köhler tried to study intelligence in chimpanzees, one of our closest animal relatives. He lived in primitive conditions on an island off the coast of Africa. He had numerous chimpanzees that had been captured in the wild when young and brought into captivity. He had cage after cage of chimpanzees lined up on one side of a long hall. Directly across from them, they kept cage after cage of chickens the scientists used for eggs and meat for themselves.

He gave the chimpanzees tasks to test their intelligence. One such task involved putting two bamboo-style fishing poles together end to end to make it longer so the chimps could use it to rake in a banana. He wanted to see how long it would take the chimps to figure out how to put one end of the pole into the other to make it long enough to reach the banana.

The chimps were sharp when it came to figuring out how to put the poles together, but when there were no bananas around to rake in, they learned to put the poles together and use them to poke the chickens in the cages across the aisle. The chickens would jump and squawk. The chimpanzees thought that was all kinds of keen.

The chimps were excited by the fact that they could get a reaction out of the chickens. But the old cliché we used to tell our children, to "ignore them, they will go away," simply no longer works—if it ever did. Bullies now use verbal put-downs like the chimps used fishing poles to verbally poke at kids who can't defend themselves. But it does no good to ignore them. A bully's behavior is motivated by the effect their bullying has on the other students—the other students laugh. So long as that happens, their behavior will never stop.

The secret for stopping bullying and name-calling is, once again, to get the students to see the harm done by such behavior; to get them on the side of the bullied kids, to get at least 20% of them to say "stop it," to change the very nature of their understanding of what has been going on in schools.

TOOLS FOR SUCCESS:
Using Films for Desensitization, Counterconditioning, and Control

Stossel's film is an incredibly valuable tool for getting students on the side of the kids who are bullied instead of on the side of the bullies. But it is far more than that; he has gone further. This film makes possible the use of two of psychology's most powerful techniques—desensitization and counter-conditioning.

The end of the film is the best. John Stossel shares his own experiences in high school. He shows a picture of himself looking and feeling insignificant, even referring to himself as a "nerd," and compares his own picture with those of the Big Men On Campus in his high school. They look like they are adults, in their thirties, even. I can't even remember any kids at my school who looked that old; they looked like they had been held back several grades. But it was very clear that the kids who do the best in school are the older students; in Stossel's words, *They were men, while we were boys.*

More than any other factor, the students who matured early had a profound advantage over those who matured late. Consider this—two kids start school at the age of six; one is six going on seven, the other just turned six. Throughout the entirety of their school career, the older kid, even though they are technically the same age, will have almost an entire year of physical growth and visual-motor coordination experience over the other. That makes them better in sports. In addition, that older kid will have almost a full year of social experience over the other. This is what makes for success in public schools. Yet, those who are at a disadvantage in age come away feeling like they are not a success; they may blame themselves and feel inferior.

Stossel's crew invited a number of his fellow high school students from his old school to the studio. Most expressed the profound feelings of, *"I just remember feeling like an outcast. I never really felt a part of any group."* Others spoke of their feelings of pain, *"My mom said 'remember this it will be the best years of your life' and I remember thinking, 'this is it?'"*
Stossel himself said of a star basketball player, *"I felt that if he would be my friend, then my life would be good."* But when they called him up, he did not even remember any of them.

All of this is of tremendous importance because, like psychologist Albert Bandura when he counter-conditioned children's fear of dogs by showing them motion pictures of a happy child happily playing with a happy puppy, it is the most effective therapy possible. First, it desensitizes teens to the feeling that they are alone; that no one else ever felt as bad as they do. Second, it counter-conditions their feelings of failure by showing them that many others have been through the same bad experiences, felt hopeless and doomed to failure in life, yet gone on to succeed. That gives them hope. It lets them know that they, too, can succeed in life. Third, if you add to that a plan to give them control over their own life, to improve their skills, like Albert Ellis, then you give them control over their future.

That is what you must get across to young people. Telling them to count their blessings, buck up, or look on the bright side is useless. Showing them that others have been through such tough times and gone on to succeed is priceless. Adding the tools that they need to succeed in life, giving them goals to achieve is what works.

THE TOOLS YOU NEED TO HELP YOU AND YOUR CHILDREN
You can watch films to help you and your children. Watch them with your children, make notes on what is important, and discuss them with your kids. Even if you are now an adult, learning that others went through the same experiences you have and still succeed, is a terribly important lesson for adults. ABC Family produces other films that provide similar, powerful experiences.
"The Wonder Years" is a marvelous series that explores so many of the problems adolescents are likely to face in life in a serious but comic look at how life looks to a teenager. Watch it with your adolescent and discuss the problems it raises.

"Teens: What Makes Them Tick" is another John Stossel 20/20 production that provides an intense view of parent-teen interaction, providing far more insight into solving problems than you will ever find in a textbook. We do not have to invent a new way of teaching reality to students; the groundwork has already been laid. Combine all these into a course for school, complete with professional comments on what these problems mean and kids commenting on how they feel after seeing their generation's problems portrayed so plainly on screen. It not only gives them insight into the problems they have, but it also helps them prepare for being parents themselves. But you cannot just do it once; you must show them examples year after year to get the ideas across. Otherwise, the power of the peer group rules.

The best work in psychology has not been done by psychologists but by the media. The reason we have failed so badly in psychology is not hard to figure out. When John Stossel does a special, he has an entire professional production crew behind him. Not just a film and sound crew, but advanced men to scout out the scene, research people to cover the basics, permissions people to get film clips from all sources, writers to flesh out the script, directors to bring it all together, editors to smooth it out, and professionals to get comments from, even a makeup crew. It takes an enormous production crew.

When psychologists make a video, it looks as if one or two psychologists make it with their spouse holding the camera in someone's kitchen with a child peeking around the corner; small wonder we have never produced anything in psychology worthy of showing to a mass audience. Most of it is not even worthy of showing to captive students who have no choice. We cannot compete with what the professional media can do.

Physicists have hundreds of billions of dollars of government money to work on expensive projects, like Super Colliders, to study particle physics. The government thinks that they may discover another atomic bomb out of the works of their creed. Psychology gets less than crumbs to study what would help our children do better in life. Yet, the power of understanding our own minds, of the role of emotion in learning, is potentially far more important for our future.

Even when professionals write a textbook for college students, typically only one or a few psychologists with little help write it. Instead, we need a professional production team, with access to a broad range of video clips, professional teams to do research, editors to put it on a DVD, gaming programmers to make it interesting, and more. This is why we need our education system to catch up with what the media can do.

To continue putting books in students' hands that no self-respecting professional could put up with is just sad. The media can do a vastly better job. But without the money and staff of the media, we cannot succeed in producing educational material that students will want to see. We need videos that get heavy ideas across to eager minds. Teachers need to be paid more, but paying teachers more to do what they do now will not change what needs to change... it is the ideas presented to students and the way those ideas are presented.

That is why we need a small-scale Manhattan project to save our educational system, but that will cost tens of millions of dollars that politicians are unwilling to put into our educational system. So, will we have to make do with more of the same? An alternative is to combine the media already available and make the best use of it to get ideas across. Parents can do this themselves, but they cannot wait for the school system to do it for them; that will not happen.

PARENTS AS PREVENTIVE PSYCHOLOGY

Parents can do something to help to prevent problems. Psychologist Julius Segal, the author of *A Child's Journey* and former director of the public information programs of the National Institute of Mental Health, describes his own traumatic childhood experiences and how critically important to him his mother's support was:

> "...As a Jewish child growing up in the 1930s in a town rife with anti-Semitism, I was never allowed to forget that I was a member of a minority group. Many of my classmates and even some of my teachers were quick to point out how atypical I was, how strange my eating preferences, how peculiar it was to run home at sundown on Friday evening to begin the Sabbath, and how out of tune I was with the realities of society. My father was a typical old-world Jew--bearded, scholarly, and strange to the mainstream of the community. I remember vividly walking along the streets of my city and becoming the object of pranks and jokes and slanderous jibes among those for whom "kikes" and "Jewboys" were the scum of the earth. I sensed myself as an outcast and a deviant, labeled with what seemed to me to be an incurable disease called "Jewishness."
>
> "I had, however, a port in the storm to which I could return each night. It was my mother who reversed the damage begun by the school society. It was she who, in subtle or brazenly overt ways, would keep driving home to me that no matter what "they" say, I was the greatest--that I had brains undreamed of by my tormenters, and potential that would one day be unleashed in more benign settings. When I was taunted at school by cruel peers or unhappy teachers, I did not fall victim to secondary deviance. I returned home to my advocate and strength. Had I lacked that support, I may well have become bitter and defenseless against the erosions in my self-concept imposed by the world outside.
>
> "My own mental health was protected by a sensitive parent who evidently knew the distinction between academic achievement and personal growth, and who recognized that no amount of learning could compensate for psychological blemishes that would threaten my sense of competence and wellbeing. Unfortunately, the same wisdom is not always found in the repertoires of those in the schools to whom we entrust our children".

Parents need to know that they can make a difference. They can ease children's pain and fears and create a positive force. This is something that should be taught in every classroom to every prospective parent.

We can take this one step further. Simply reading stories like that of Julius Segal is not just informative but also helps make us realize how arbitrary and unimportant others' value judgments are. It helps to counter-condition us to the fears embedded in youth's brains. And some of what is in this book is in the same vein. When we learn how very similar the fears, depression, and anxieties are that other people experience, it makes our own less fearful. It makes us feel less alone and better able to put our own problems into perspective.

One of the most successful American executives is Lee Iacocca, former CEO of Ford, and the man credited with single-handedly bringing Chrysler out of bankruptcy. He tells of his own childhood experiences in which other children singled him out for being Italian in school, calling him a "wop" and a "dago" and unmercifully teased him about eating the then-new "pizza pie." It does not sound like much, but it is a memory that stayed with him for more than fifty years when so many other memories were lost. He remembered it for that long and wrote about it fifty years later in his autobiography as one of his most significant memories.

Iacocca learned something of what the world would be like from his childhood experiences. The world is not fair. Yet, through it all, he had the support of his family, especially his father. Iacocca described it:

> *"Whenever times were tough in our family, it was my father who kept our spirits up. No matter what happened, he was always there for us. He was a philosopher, full of little sayings and homilies about the ways of the world. His favorite theme was that life has its ups and downs and that each person has to come to terms with his own share of misery. "You've got to accept a little sorrow in life," he'd tell me when I was upset about a bad grade in school or some other disappointment. "You'll never really know what happiness is unless you have something to compare it to."*

The problems Segal and Iacocca describe have nothing to do with being Jewish or Italian; they are the same problems faced by any child with a strange name, who looks or acts funny, who is not "cool and popular," who is the last to be chosen in gym class, or who simply makes a mistake in full view of others. It is all part of the laughter and finger-pointing, the bias and belittlement that greet anyone who is different.

There is some degree of irony in a similar story by one of the greats in the history of psychology, Phillip Zimbardo. Zimbardo tells of how, as a child, he had to find new ways home from school to avoid the abuse by groups of youths who would torment him daily for being Jewish. The irony, he says, is that he was not Jewish; he was Italian. But that mattered not one bit to the youths; anything that makes you look different makes you a target.

Of course, there was nothing wrong with Julius Segal, Lee Iacocca, or Phillip Zimbardo; they were all quite good people. But they had been labeled, as if they had cancer, by other people's words.

Anything that makes you stand out makes you a target for laughter, derision, and name-calling. If your name is strange or cute, you become a target for ridicule. If you make a mistake in front of the class you become an object of laughter. If you have cerebral palsy or mental retardation, you become the subject of finger-pointing and whispers. If your sexual behavior is different, you become an object to be discussed behind your back. If you believe in a different version of truth, then you become a target for anger or ridicule. Overweight people become a target for comedians looking for a quick laugh. If your race or religion is different, you become a target for suspicion and jokes.

Racial and religious prejudices are not unique; they are just two more things that make people different. In one way or another, we are all victims of other people's judgments. And it is all prejudice. We no longer tolerate racial prejudice in America, so why would we continue to tolerate the kind of prejudice that has done such harm to our children and ourselves?

It is not inevitable; it can be changed. But it can only be changed by educating people to understand their own minds and the harm they inflict on others with their value judgments. It is a lesson that must begin in grade school.

If there is no one to support us at that early age, then we may come to see ourselves as others do. Children or teens may develop a feeling of "learned helplessness," a feeling that there is nothing we can do. Or, we may go on to develop self-contempt out of the feeling that we have allowed ourselves to be taunted, used, abused, and pushed around by others. We may feel that somehow, we should have been smart enough to have done better; that what happened to us was somehow "our fault."

The greatest tragedy is that, instead of recognizing this as the "insanity" of our society for allowing it to continue, we blame it on the individual. We make it "their" problem. We invent labels in the DSM-V manual to tag their problem... as if inventing a label will somehow explain it. Or we give them medicine as if a pill will solve their problem.

For all the years of recognizing the problem, we have done nothing to change it. Some children grow up to be depressed or fearful. Some experiment with drugs out of curiosity or to ease the pain of living. Others lash out in anger. We blame them. We make them criminals. If their thoughts are sufficiently unique, we get out the Diagnostic Manual and tag them with a "mentally ill," "dysthymic," "bipolar," or "oppositional defiant." We reduce their problems to a label. Then no one bothers to deal with what really causes the problems.

We tell people they must be held responsible for their behavior. Yet we do nothing to change the far greater crime; the fact that we allow it to continue. There is no one to hold us responsible for our failure as a society.

THE GREAT CESSPOOL OF PSYCHOLOGICAL TRAUMA: SCHOOL

The school system has been a breeding ground for psychological problems, just as surely as the floods that caused the cesspools to overflow into drinking water was the breeding ground for cholera in hotspots along the River Thames. It is not that public schools are inherently bad, but because we have ignored the very real problems they create.

Only within the last one hundred and fifty years have we created a prison for our children where they must go for twelve years. We dump our kids into the school system and let them sink or swim. Some do very well; some do very poorly; most just muddle through.

It is not just a matter of helping those who need help; the majority of students come out of the school system feeling like they are not a success at anything because they aren't. The only two things we give them they can try to succeed at are sports and socializing; nothing else is valued by the other students.

If public schools have been a cause of psychological problems, they can also be a cure. We can educate our children to be sensitive to other people's feelings, be aware of how other people's behavior affects us, learn to avoid problems created by other people's words, and know what to expect from the real world. But we do not.

Parents need to know that they can make a difference, but it is not sufficient. What is needed is a drastic revision of our educational system. No single need for change in the public schools can be more important than this; schools should have a responsibility to teach children to be sensitive to the harm they do to others... to understand the importance of their responsibility for helping others... to understand their own minds and those of others. So long as we teach only reading, writing, and arithmetic, we fail to teach the most important lessons of life.

PREVENTING PROBLEMS THROUGH EDUCATION, NOT THERAPY

Preventing problems is the key to the future of psychology. Education, not therapy, is the "cure" we must pursue. We will not solve our problems by finding a new chemical to sedate our fears or a new therapy to root out the pain of living. We need to educate ourselves as to the causes of psychological problems so that we can avoid or change the problems before they begin... in the same way that we prevented plagues, cholera, malaria, and—eventually, we hope—cancer, by changing the forces in the *environment* that cause the problems.

When the World Health Organization or the National Science Foundation discusses prevention, the tendency is to first identify those kids who need help in schools, which means signaling them out and labeling them, and second providing them with intervention in the form of therapy or pills. In addition, for those who were singled out for help, they hope to provide ways of giving support to the family and individual of those identified as "at risk."

All of that sounds quite logical, but it is absolutely not what I am suggesting here. The very last thing we should do is to be in the business of labeling kids and providing more work for those who profit from their problems. Instead, we should concentrate on providing the education, knowledge, desensitization, and counter-conditioning that everyone needs to make it through life and the tools they need to gain control of their own minds.

THE KEY LESSON OF PSYCHOLOGY:
We Have Seen The Cause of Human Problems and it Is Us

In case you have not noticed, the number one cause of most human problems is not genetics, biochemical, or serotonin in our fevered brain; the number one cause of most of our own problems is other people's value judgments. In ordinary human interactions, value judgments cause the vast majority of our human problems. Sometimes, we are the ones causing problems for others by our value judgments.

But despite that, the overwhelming causes of human psychological problems are the emotions embedded in our brains by people—your parents, peers, lovers, boss, associates, and yourself.

Even in biological problems such as schizophrenia and ADHD, other people cause problems. People with these problems behave quite differently from others. So, what do other people do to those who are different? We belittle and humiliate them, talk about them behind their back, ostracize and ignore them, and put them down. We create vastly greater problems for anyone who is different.

And the overwhelming cure for most problems has to be in developing social skills to deal with them, learning the knowledge needed to understand those problems, and getting across to others the need for understanding and developing an ability to control the forces in your own mind.

PEER GROUP PRESSURE AND BEHAVIOR

Studies show that even police officers are more likely to abuse your rights in front of other police officers than if they are alone, showing off to their fellow officers how tough they are is a powerful incentive when you have an audience. And no officer ever tattles on a buddy, or they risk being hated by their fellows.

You see the same fear in Congress when Republicans were asked to criticize President Trump. No one wanted to be a tattletale, someone who "rats" on a buddy. Their silence was deafening. But it all begins in the school system with the learned fear of being labeled as a "tattletale".

Congress sees nothing wrong with taking in obscene amounts of money from lobbyists for their reelection campaign and then voting the way the lobbyists for the drug and insurance industry, big oil, or whoever "donates" the money expect them to vote. "Everyone else does it," they say, "so we have to do it too." That must make it OK.

"If everyone else jumped off a building, would you?"

"Duh, yeah!"

And automobile accidents? You hear in the press of old people mistaking the gas pedal for the brake and running into a crowd of people. This sensationalism pandered to by the media, led to laws being proposed that would force the elderly to undergo yearly driving tests to certify if they are still fit to drive. Yet, the statistics are staggering—old people have one of the lowest accident rates of all. Teens have five times the accident and fatality rate as the elderly. This is not new information; insurance rates have always been much higher for those under 25. Yet the press never informed the people about this; instead, they opted for sensational accounts of the elderly blindly plowing into crowds, and that accounts for the bills—one of which is now before the Texas state senate in Austin that would require routine driving tests for the elderly.

Why do teens and young adults have five times the accident rate as any adult over the age of 25? Some say it is because their brains are not yet mature, or it's raging hormones, or maybe even that they feel invincible. But the most likely explanation is found simply in watching what teens do when they ride around in a car. If they are with a bunch of other teens, they are all talking, not always paying attention to their driving. Or they are texting or talking on their cell. Simple psychological factors involving attention account for most accidents, not hormones or an immature brain.

At the dawn of the twentieth century, you could still buy the ingredients of medicines "miracle drug" *laudanum* from any corner drugstore. Laudanum was a mixture of opium and alcohol and was once widely hailed by doctors as a miracle cure for every imagined disease. Thousands of people wrote testimonials to its curative power. In fact, it cured nothing at all, but it was said that after two or three drinks, you didn't much give a care what you had.

If that did not cure you, Sears Roebuck's Department of Electric Belts would sell you the newest medical marvel, a Heidelberg Electric Belt, guaranteed to cure cancer, impotence, and stomach diseases, by the latest medical marvel, electricity. A loop hung down over the front of the private parts, a veiled hint at its rejuvenating sexual power. It was the Viagra of the day, and Sears promised to send it in plain brown wrapping to avoid embarrassment to the user. It came complete with voluminous testimonials from satisfied users who swore by its mysterious power to cure "any disease" —from menstrual pain to, yes, that dreaded disease, "cancer of the stomach." And, of course, Sears' famous money-back guarantee—if you died, they would give you your money back. Out of the thousands of letters from satisfied users, not a single dead person ever asked for their money back.

People like to read stories of the "amazing hidden secret" powers of the mind. Yet, the real testimony of history, with its thousands of years of bloodletting, electric belts, witch hunts, wars to end all wars, and our absolute certainty that the earth was flat and in the center of the universe, all led us to a different conclusion. We like to hear stories of the amazing human mind and its secrets and extraordinary gifts. But if you consider our real past as a species, the endless wars, witch hunts, bloodletting, and weapons of mass destruction in Iraq... then that is not what history has to say about us. The really important lesson of history is that we are all incredibly stupid.

That is not what people want to hear, of course, and the media often has an educational program talking about how much smarter we are than a chimpanzee or how much more civilization has improved with our brain than our primitive relatives. Yet chimpanzees never spent thousands of years torturing their grandmothers into confessing they were witches and burning them alive at the stake.

Times slowly changed. We learned to prevent drought by irrigating our crops and use crop rotation to avoid soil depletion. The plagues that decimated half of Europe we never cured, but we learned to prevent many of them simply by improving sanitation. Burning our grandmothers did not prove useful. Today, medicine explains the mind as genetics or biochemistry. And psychology and psychiatry still search for a magic spell; a chemical, poultice, or mustard plaster; a cathartic bloodletting... a new therapy or a drug to scourge the demons in our minds and heal the pain of living.

We have failed to grasp the most profound lesson of tens of thousands of years of magic and medicine; that the most useful tool of medicine is not drugs but preventing the problems in the first place. We did not cure the epidemics of cholera or typhus by drugs... instead, we learned its cause from epidemiologists like John Snow and others who tracked cholera outbreaks and found that they originated in a certain season of the year; the rainy season. They then found that the epidemics were heavily concentrated around water in wells near the River Thames in London. Putting two ideas together, they realized that the river overflowed during the rainy season, forcing effluent and feces up from the outhouses into the wells where they obtained their drinking water.

Eventually, they realized that, even though they could not cure the disease or even see the microorganisms that caused it, they could prevent the disease by changing *the environment;* by boiling the water to kill the organisms that caused the diseases. We helped prevent epidemics that killed millions by simply improving sanitation. They were never able to cure the disease by medicine. Even today, following a local river flooding, we warn people to boil their water before drinking.

We helped prevent malaria by eliminating mosquitoes in *the environment* that spread the disease by draining swamps or using insecticide. We can help prevent heart disease and stroke by controlling high blood pressure—a major factor in these diseases—by reducing cholesterol levels and changing diet. All this is far more effective than trying to treat the diseases after the damage has occurred.

PSYCHOLOGICAL PROBLEMS BEGIN IN THE ENVIRONMENT, NOT OUR GENES.

Even crime has a psychogenic origin. Teens commit three times more crimes than adults. This is often blown off as just teenage energy or teens trying to "find themselves," yet there is a profound psychological reason for this as well. Teens have four times the unemployment rate as adults. When you have no job to make money, and you are no longer in school, what do you do with all that free time? You ride around with your friends, your peer group. And we have known for a very long time in psychology that teens and adults will do incredibly stupid things in a group they would never do by themselves.

So, one teen in the car says, "Hey, let's go paint graffiti on the school walls!" Or another says, "Let's break into an ATM machine." The other teens, often afraid to look chicken in front of the others, say, "Yeah! Cool!" and they end up with a felony arrest and conviction. Maybe that is a step up from drowning, but it is all caused by the same psychological forces; emotions embedded in the mind by the conditioning of our peers, by the fear that others would think we are "chicken."

For adults behaving badly in social groups, you need only look to the mistreatment of prisoners at Abu Graib, where American soldiers stripped them naked, forced them into "sexually compromising" situations, and turned vicious guard dogs on them. Or the massacres by American soldiers of 450 old men, women, and babies at the villages of Mai Lai and Mai Khe in Vietnam.

Only a handful of the American soldiers did the killing, but of the overwhelming number who did not, not one tried to stop it. Not one spoke out against it. Not one turned in their buddies. It became public two years later only after the photographs of army photographer Ron Haeberle was made public. Then no one could deny what had happened.

Of course, we could easily teach this to our children in public schools. We immunize them against the effects of group pressure. Every grade could be shown examples of what to expect from peer group pressure. Society being what it is, such changes are not going to occur in time to help our children, much less you or me. Which brings us to the next point: What do you do for the rest of us?

The fear of what people will think is a conditioned emotional reaction. It differs not in the least from any other phobia. It has the same basic origin as the fear of cockroaches, unfriendly dogs, the number 13, or Iraqi weapons of mass destruction (remember that one?). Our entire nation was told they had massive stockpiles of sarin gas, mobile biological warfare facilities, and were only a short time away from developing the atomic bomb. We were so afraid, that we invaded Iraq, starting a war that took the lives of over 92,000 Iraqi civilians, according to U.N. figures. But we found no weapons of mass destruction whatsoever. And we never even apologized. Being number one means never having to say you are sorry.

Like all phobias, the fear of what other people will think can be counter conditioned without invading anyone.

Only a short time ago, a client where I worked became suicidal and uncontrollable. An attempt was made to place her into an intensive therapy program at a beautiful new psychiatric hospital that had opened nearby. We were informed that the fee would be $900 a day. Nine hundred dollars a *day*? I know some of the professionals who work there, and I am sorry, but there is absolutely nothing they do that is worth nine hundred dollars a day. Who could pay that kind of money? They can only get away with it by charging it to insurance companies, the government, or the rich. Unfortunately, that has become the trend in American healthcare. And the treatment they get... drugs, talk, and a two-star class motel room.

The people who own such institutions now pay for their palatial estates at the expense of the people who are hurting. We can do far more by preventing such problems in the first place... by education rather than by therapy; by training parents how to be decent, capable, caring individuals; by teaching children early on how to avoid such problems; by providing understanding. If we wait until the individual is an adult, until the problem has exploded, then it is much more difficult to deal with.

speaker was considered to be a rather "warm" person, while the other half read that the speaker was considered "cold."

After all had heard the same speaker, Kelly asked them to turn over the biographical sketch and complete a rating scale about how they felt about the speaker. Those who received the sketch with the rather "warm" description rated the speaker as more likable, a better speaker, and someone they would like to meet than did those who received the "cold" description.

Some of the students may not have even paid attention to the description, yet the fact that a single word could change their perception of another individual at all is remarkable.

We all know from our high school experiences that rumors we hear about another person can change the way we react to that person. We may insist that we never pay attention to rumors, yet the first time we meet that person after hearing a rumor, the very first thought that pops into our brain is that rumor. This has a dramatic effect on who we like or dislike, on how we react when talking to that person, and on who we vote for or against in politics—even on how we judge ourselves. We all believe we know enough that our mind would not be tricked by something so trivial, yet the evidence suggests we do not understand our own mind at all. Recent political anger in America shows how easily emotional conditioning can control the mind.

Our brain's very biology is programmed to make value judgments. The center of human and animal emotions is the limbic system, which tags experiences with emotion. Part of that system is the amygdala, the center of fear and anger (the "fight or flight" response). It is a bulbous mass right near the end of the hippocampus—the part of the brain that makes memory permanent. An emotional experience is more likely to be made into permanent memory.

Every emotional experience is tagged with a positive or negative emotion—a conditioned or learned emotional reaction. This is the origin of our value judgments. From Voodoo to cancer, from right to wrong, from whom we like to whom we dislike, words tagged with feeling trigger emotions in our brain. We come to judge others by the emotions these stimuli trigger in our brain.

WE DO NOT REACT WELL TO ANYONE WHO IS DIFFERENT

It exercises its power with such a light touch, like the fog of Robert Frost, "... on little cat feet." How is it possible that a force such as our fear of "what other people will think" could be so delicate, so indirect that we are unaware of its impact?

How could this same power be so potent that it could force one of America's greatest authors to censor his own words?

At the same time, how could such a subtle force be so compelling that it could force a person to take his own life?

You may think it exaggerated. Surely, no fear of what other people think could so torture the mind that death would seem preferable. Yet, 5,000 teenagers and young adults in America take their own life every year. Their reasons may differ, but it does not happen without cause, and it is a profound comment on an enormous degree of unhappiness in our culture.

The prime causes are feelings of hopelessness, failure, and isolation, most of which are engendered by fearing what others will think, or fear attached to a label like "failure," which evokes an emotional response as great as the label of "cancer."

There is no biological cause or genetic code that forces us to shape our lives to other people's ideas. No tyrant demands that we conform or die. The tyranny has been embedded in our brain. The fear of what others will think is forced upon the mind early, forged hot in the verbal fires of childhood. It begins when children call each other names, like "fatty," "stinky," or "stupid."

Children learn to avoid the sting of being singled out by the laughter of others. They fear making a mistake, being ridiculed or thought "stupid" or looking "silly."

We learned to fear it from other people's warnings and ridicules. . . parents tell us "don't act silly," teachers tell us to "sit down and act right," but, most traumatically, it comes from our fellow students' taunts and laughter. It begins in our school playgrounds. Our school system teaches children this hard lesson at an early age, without ever intending to. It is one lesson our schools teach well.

On January 2, 2009, Judith Scruggs, a single mother working two jobs to support her children, could not find her son, 12-year-old Daniel. Daniel was "a typical 12-year-old," very smart but with a learning disability. His mother knew little of what his life in school was like. Later, other students shared how he was bullied every day in school. Rebecca Leung of CBS would later interview other students who described how he was hit, kicked, and spit on, and once made to eat his lunch off the floor by the worst bullies. When his mother went to his room that day, she found Daniel had hung himself in his closet, dead at the age of 12.

The police declined to charge any of the boys who had bullied Daniel or anyone in the school system who had been unable to protect him. Instead, they charged his mother with negligence. The school said they had recommended the mother get help for Daniel nine months earlier. But there is little psychotherapy could have done to help Daniel, as long as he had to return to the bullying at school.

To insist that Daniel is the one who needs help is to ignore the real problem and source of pain that led him to commit suicide. And they blamed his mother for this? It begs the question, why was the school system unable to help? Why did others try to blame Daniel's mother for the system's failure? Would it even have helped if Daniel got counseling and then had to go back to the daily bullying? Not if he had to go back to the environment that created the problem. We have seen many children who got "counseling" that did no good.

In April of the same year, eleven-year-old Carl Walker-Hoover, a junior at New Leadership Charter School in Massachusetts, committed suicide after months of bullying that the school had been unable to stop. Carl had repeatedly been subject to name-calling and taunts, often calling him a "fag" or "gay." But Carl was not gay; he was eleven years old. He had no testosterone. Carl never even knew what the word meant—he had to ask his mother. Carl committed suicide at the age of eleven.

To their credit, CBS's "60 Minutes II" did a special on the suicides, and other networks did cover them, but then everyone dropped the ball again. It disappeared from the news to be replaced by "balloon boy," who supposedly attached balloons to a chair and was swept into the sky. This received far more coverage on all the networks. Even after it turned out to be a bogus story, the media went on and on about this trivial fraud.

Little boys may strut, posture, and brag to others—not so much in pride, but as in an effort to avoid being different. They react to taunts of "chicken," "yella," or "scardy cat" with gestures of defiance and false bravado. They brag or take a dare as if to say "I am worth something!" They even go to the opposite extreme of trying to prove they are macho by doing dangerous things they think others will admire.

THE PERILS OF PAULINE: ADOLESCENCE

The uncertain years of adolescence bring dramatically increased fears of what other people will think. The fears that started early become more intense because society makes it so—not that everyone experiences the same fears. Yet. at one time or another, we all do.

Anthropologists Gary Schwartz and Don Merten vividly describe the primitive and savage initiation rites of adolescence in a contemporary American high school in California. Their study involved the ritual cycle of initiation into an elite girl's sorority.

The process of selecting the elite members who would be allowed to join the sorority (as opposed to the "nerds" who would not) was set against a firm democratic ideal. All the girls expressed their absolute belief that everyone is equal, and no one is intrinsically superior to anyone else. It was the classic American ideal of social equality: Thomas Jefferson in petticoats.

Yet, only moments later, they began to decide who was worthy of being in their elite sorority. In sharp contrast to their democratic ideals were personal traits that they said would disqualify other girls from joining their group. Those who were not "cool" need not apply. Those who were dull, ugly, shy, unenthusiastic, and so on would not be accepted.

The cycle began with *rush* invitations. One girl described the fears and anxieties she went through, waiting to join in the hope of getting an invitation. One by one, all her friends received invitations. She did not. In typical American fashion, it had a happy ending—her invitation had been mailed to the wrong address. Many were not so lucky.

After the rush party came the blackballing. In a secret meeting, the girls would get together to pass judgment on the good and bad points of the personalities of those who had been pledged. Were they worthy of being in "our" sorority? Those who were selected were expected to be "cool but not too cool, sweet but not saccharine, deferential but not obsequious, gregarious but not self-assertive, demure but not shy."

Mostly, they were expected to be both "*cool*" and "*popular*."

Those who were allowed to pledge went through an eight-week period in which they had to curtsy and greet other members with "Good morning, your royal highness." They were expected to perform menial tasks for the members, and they were often made to do foolish and childish stunts, such as asking boys to marry them and proposing to a mailbox. During the *Yelling*, members would scream at pledges for their failures; pledges would be accused of letting the sorority down, and be told they were unworthy in some way.

On *Hell Night,* the harassment increases. The focus is on traits that members of the sorority think would make their sorority less elite. Pledges are asked, "Why do you date that creep? Why are your legs skinny? Will you do anything for your sorority?" Finally, they are told to cry on cue. If they do not or cannot, the other members scream at the pledge until she does. If she still cannot, then her pledge bows are burned, and she is stricken from the club. Sort of like being voted off by the tribe on the Survivor series on TV, as they extinguish your flame, the host says, "The tribe has spoken."

This was in high school?

Certainly, none of this is limited to high school. I have known college students who were psychologically devastated and eventually dropped out of school because their friends made it into a sorority, but they were blackballed.

Three young women at the university were friends. They were all outgoing, friendly, and enthusiastic about life. All three pledged the same sorority; one was blackballed. Like the man cursed by voodoo, she changed overnight from being a happy and outgoing person to someone who walked around with her head down, never looking up and avoiding her friends because a sorority rejected her; the same sorority that took two of her friends. She felt worthless, all the more so because her friends got in and she did not.

Why was she rejected? Because she did not believe that the sorority expected her to take seriously all the silliness she was required to do. They did. They considered her unworthy because she did not take it seriously enough. After a semester of depression and avoiding her former friends, she dropped out of school, never to return, rather than go through another semester with the feeling of rejection.

Some psychologists might call this an identity crisis. Others might say it is a matter of an ego-devaluating experience. But it all boils down to one simple fact— she was stamped with the emotional mark of a "failure," and she was ashamed that her former friends, their roommates, would think less of her. She had been labeled as if she had been told she had cancer. It was our culture's version of the voodoo curse. She was afraid of what other people would *think*.

It was not an entirely irrational fear. She was quite right—other people do think that way. We have not taught our children any better. They have not learned the relativity of their value judgments, and we have failed to teach an understanding of the harm done by our value judgments.

The same fears occur in males who pledge fraternities, although they are not as likely to talk about it in public. Former Dallas Cowboys' football star Peter Gent was a wide receiver at the height of his career; he knew the roar of the Sunday crowd as he caught the passes of Don Meredith and scored in the end zone. He played with pain for the good of the team. After three years, he was injured. He was cut; it was all over. No more cheers; no more glory.

Gent later wrote a novel, a best-seller, *North Dallas Forty*, where he vividly described the feelings and emotions that surround the individual caught up in the social order of our society. They made a Hollywood movie out of the book, but it was never as good. It is an extraordinary book on male social bonding and the reality of fame and football. But it is not just a comment on fraternities; Gent clearly meant it to apply to the glory and fame of football, where players are expected to play with pain and sacrifice themselves "for the good of the team."

Now, crazed on Mescaline, driving across another college campus twelve hundred miles and several light-years removed, I was beginning to understand. If a man is lonely enough, he will eat raw eggs, carry olives in his ass hole, and let homosexual history majors from Flint beat his butt bloody with a paddle. He does it all in the belief that with the new morning they will have learned to love him by brutalizing him.

But when the ritualized humiliation ends, how can he admit to himself that it had no meaning, that he is still alone, only momentarily distracted from the fear and loneliness and hatred that consumes us all?

It is one of the tragic ironies of our culture that the "fear and loneliness and hatred" that has its origin in our youth experiences continue to be a major force in many people's lives. It differs in degree from one person to the next; the majority do not experience it so intensely, and most are unaware of its effect on their lives. Few learn how to escape it.

PREVENTING THE PAIN:
Education as Preventive Psychology

None of this is inevitable. We could teach young people what to expect before it happens just by telling them stories such as ones about girls in a sorority. This would give them the ability to decide if they wanted to join a fraternity or sorority, and it would give them an understanding of what to expect if they do decide to join.

"...were not hosting an intergalactic kegger down here."

Men in Black

Simply by telling students actual stories, such as the one about the girl who was rejected by a sorority, would allow them to mentally steel themselves for the possibility of rejection.

Students steel themselves for the possibility they may flunk a test by thinking, "I probably blew it on that exam," even if they think they did well because we have learned from past experiences that we do not always do as well as we expect.

In addition, they need to know just how fallible the value judgments that others pass on them really are. That is good preventive psychology; it minimizes the danger of emotional depression.

Every year in America, students die in fraternity hazings. Much of the time, it is due to being asked to do something stupid as part of the initiation, like chug-a-luging alcohol. You might get away with chug-a-luging beer, as you may pass out or throw up before it kills you, but hard liquor is a different matter. You can ingest enough to make you pass out, and your stomach will continue to absorb it.

One fraternity initiation at Southern Methodist University happened only two weeks after all the fraternities and sororities were called in and told that hazing would not be allowed. They went ahead and had a pledge upend a bottle of hard liquor. He passed out. Instead of calling an ambulance, because they were afraid of the consequences if they were found out, they tried to walk him around for three hours without success.

Finally, they called an ambulance. He lived, so he did not become a statistic, but he suffered brain damage from the incident. You would think warnings would be enough. No. You have to show them examples such as these to get the point across. Education in advance can be far more valuable than therapy after the fact. Yet, all we give our youth are stories of success, of sports heroes and cheerleaders, while censoring and ignoring the pain, heartache, and frustration of life.

Maybe we are hosting an intergalactic kegger.

History books tell only of success and accomplishments. Columbus discovered America; so, we named a holiday after him. Columbus died in debtor's prison. Cortés conquered the Aztecs and gave the Indians the gift of smallpox. The Indians, in turn, gave us syphilis. That, we do not mention.

Few of us will ever suffer the pain of a Lincoln or Churchill, who had to live with the knowledge that others died from their decisions. Yet, we will all feel such pain at some point in our lives.

The human mind does not readily distinguish between big and small pain. The limbic system reacts to the same degree to a teen being jilted by a lover as it does to a serious problem of losing a loved one. The terrible pain felt by Abraham Lincoln over sending young people to their death in war was no greater than that of a teen who is bullied in school. The teenager who is put down and called names by others may feel as intense a psychological pain as Churchill, who was fired in humiliation after the battle of Gallipoli that he championed. The emotions are the same. Even with time, it is often hard for us to put reality into perspective; the emotions are so real to us.

Nowhere in our schools is the slightest hint of the kind of problems faced by the "greats" of history. Nowhere is there a hint that others have felt the same. This is the reality that is left out of our schools, censored by the fear of angering those who want our schools to glorify our history, to make our history as devoid of honesty as a Hollywood movie.

The totality of our cultural glorification of success and love can be judged by the fact that it is so difficult for young people to put into words what their fears and frustrations are. Instead, they end up saying, "I don't care," trying drugs to kill the pain, or becoming angry at parents because they do not understand their feelings of anomie, loneliness, or despair.

What young people need are goals for the future, social skills for success, honest information about what they can achieve and how, and adults who give them a way out or who reduce their fears by telling them about their fears when they were young. They need skills, direction, knowledge, and hope. That is not what they get. Instead, adults react with the platitudes they heard when they were children. They talk about how much better off the younger generation are than when they were young and how they should appreciate what they have.

The young learn to avoid even trying to explain.

KNOWLEDGE IS THERAPY FOR ADULTS AS WELL

Adults do not hoe an easier row; the superstitions and fears of youth merely change their color. The eighteenth pin in the voodoo doll still pricks the mind. The illusions seem so real. No one seems to grasp the nature of those fears.

This may be one of the great strengths in a book about the mid-life crisis by Gail Sheehy called *Passages*—not because she grasped any great underlying scientific principle, but because she had the courage and ability to spell out, in no uncertain terms, those terrible fears of failure; the confrontation with being alone that so many people feel but can never turn into words.

Gail Sheehy could put those feelings into words that everyone could look at and say, "Hey, I know that feeling. I've lived with that. I'm not alone. By God, it's even common!"

We cannot peddle placebos and panaceas, as so many others have. We must simply give people a glimpse of what reality is all about. That helps relieve their fear of society's verdict and altered the measure of their lives. It put their lives into perspective.

We have to let people know they are not alone, they are not "sick" like Holden Caulfield, they are not a victim of a biochemical rage in their brain, their genes have not mutated, and their serotonin has not tanked; moreover, they are not a failure in life—there is hope.

That kind of knowledge, that simple understanding, often has done more to ease their fear of the eighteenth pin than all the lithium levels and psychoanalytic "talking cures" have been able to do. It was the first small step toward giving them hope.

The next step is to give them control over their own minds.

The evidence is overwhelming that fears, superstitions, and traumas common to us all are forced upon us by our common experiences, growing up in our society, and our families and schools. Moreover, we can and must change the problems at their source: in our child-rearing practices and school system.

OUR SCHOOLS:
A Failure to Educate

Former First Lady Michelle Obama recently wrote in her book about the feelings of shame she experienced following a miscarriage. She is hardly alone. Our school system has failed to educate our youth on what to expect from life and how common miscarriages and spontaneous abortions are. Instead, we allow them to go into life unaware of reality and into feelings that somehow, the problems they face are their fault.

More recently, Megan Markle, who married the British Royal, Prince Harry spoke of the psychological pain of having a miscarriage. A newswoman talked about the terrible feelings after having a miscarriage. Yet no one talked about how knowledge of how incredibly common miscarriage are, instead, we let women think that babies are a "miracle" and often warned about what we eat, drink or smoke can cause birth defects. We study birth defects in college classes. Yet few ever learn that only 31% of all conceptions result in a living newborn baby. When they have a miscarriage, they often feel guilt, as if it was caused by something they did, instead of realizing how common it is.

One of the better examples of how simple education could prevent psychological problems is found in an autobiography by model and actress Brooke Shields, *Down Came The Rain*. She tells of her own experience with depression that left her with suicidal thoughts after the birth of her baby. My apologies to her in advance for adding my interpretation, but it is a beautiful story that needs to be told. Moreover, the lessons from this story relate to all forms of disappointment and depression.

Brooke Shields has had an incredibly successful career as a model and an equally successful career as an actress with roles in *Pretty Baby, The Blue Lagoon*, and even a television series *Suddenly Susan*. She has millions of dollars and a fancy estate worth millions with a Mercedes in the driveway. And here she is going into severe depression over not being able to have a baby? Some of my students have said that they would be happy to trade one or two of their children for one of her Mercedes (joking or not). Most of us would be thrilled to have a fraction of her success or money, but success and fame do not bring freedom from worry or depression.

First, she took time off from acting to concentrate on having a baby, but she could not. She described herself as *"feeling like I was failing at everything."* Then she had seven in-vitro fertilization. She describes how she would get her hopes up, and they would be dashed every time. She describes feelings of guilt: *"Why am I being punished, what have I done..."* Each time, she would try again, and again, get her hopes up only to be disappointed. Finally, fertilization took. She was pregnant—a joyous experience! But then she miscarried. Her hopes were again dashed. She went into feelings of guilt, *"Why me? Why am I being punished?*

EXPECTANCY EFFECTS:
The Elation-Depression Effect

The Elation-Depression Effect is an important lesson we will all experience. This came from a behavioral psychologist named Tinklepaugh, a great name for a psychologist—other kids had a field day making fun of his name. Tinklepaugh was working on his dissertation about memory in chimpanzees. He would show a chimpanzee a piece of lettuce and let him watch as he placed the lettuce under one of three flowerpots turned upside down. He would then put a barrier between the chimp and the flowerpots. He wanted to see just how long a chimpanzee could go and still remember which pot the lettuce was under. Chimps have incredible memories, and no matter how long he made the chimps wait, they ran immediately to the correct pot and got the lettuce.

Sometimes, Tinklepaugh would use a banana instead of lettuce. The chimps loved the banana, but they worked quite well for the lettuce. One day, when Tinklepaugh became bored with his experiment, he decided to pull a trick on the chimp. He showed the chimp a banana and placed it under a flowerpot. Then he put the barrier down so the chimp could not see, and he replaced the banana with a piece of lettuce.

After the time passed, the chimp came running over, lifted up the flowerpot expecting a banana, and found a cruddy piece of lettuce. The chimp shrieked. He became exasperated and refused to continue. Legend has it that he threw the flowerpots at Tinklepaugh.

The really important point of this story is that the chimp had worked quite well for lettuce before, but now that he *expected* a better reward, the banana, the lettuce he liked previously had become something very negative.

You see the same behavior in students. Two students take the same exam, and both get a C grade. One is elated, the other is depressed. Why? Both students got exactly the same grade. How could they possibly react so differently?

The first student is afraid he/she failed the exam, so getting a C is a reprieve. They still have a chance.

The second student expects an A on the exam, so getting a C is like a slap in the face.

Most students in my classes are going into nursing or medical programs, and psychology is required before they can continue. More than this, they must have an A or B to be accepted into the medical program. So, if one gets a C on an exam, you can often see the blood drain from their face. It is psychological, like a voodoo curse, yet it is as real as a slap in the face.

So what does this have to do with depression? Getting a C when you are expecting an A truly is a depressing experience. Ask the students. They sometimes feel physically ill—enough to want to drop the course. Some of this anxiety and depression can be relieved simply by letting them know that they can earn extra credit or even take the class over again if necessary.

In 2008, the American real estate market collapsed, and the banking industry began failing. Hundreds of billions of taxpayer dollars were used to prop up AIG, Citibank, Bank of America, and more to prevent a repeat of The Great Depression of 1929. Then, Bernie Madoff's Ponzi scheme also collapsed. Investors had put over 65 billion dollars into his fraudulent investment firm. Madoff was arrested, but there was little money to pay off his investors. One man, who had been a billionaire, lost almost everything; he was then only a millionaire. He committed suicide by stepping in front of a train.

Most of us would be happy to be only a millionaire. To him, it was such a loss, such a mark of failure, that he could not endure it. No one had prepared him to understand the pain he would feel; no one taught him what to expect from his own mind.

But what does this have to do with Brooke Shields' postpartum depression? Everything. What would happen if a woman were to bring her newborn baby into work or class? What would all the other women do? They would all run to see the newborn, all with appropriate "Oooohs," and "Awwws," and "Coos." And statements like, "He/she's so precious," "Isn't he/she just adorable!" "Coochie, coochie, coo!" Or whatever women say. And there would be lots of good feelings all around.

Yet, that is only 1% of the time your newborn baby takes. The rest of your time is taken up with feeding, burping, and wiping. If the "ooooh" and "aaaaw" and good feelings are your only image of what having a baby is like, if you are unaware of what to expect and have unrealistic expectations, then this is like expecting an A and getting poop. The jarring reality can be difficult to accept for those who have unrealistic expectations. That is one reason why we need to do a much better job of educating young people on what to expect in real life, not letting other kids or ourselves fill their minds with unrealistic expectations.

When Brooke's baby was finally born, she describes her first reaction as, "Where are the birds singing? Where is the fairy dust?" She seems to have had an unrealistic expectation that led to an inevitable disappointment. More likely, this may be a combination of all the elation-depression effects—repeatedly getting ones hopes up after in-vitro fertilization, only to have her hopes dashed, getting pregnant only to have a miscarriage, along with the elation-depression effect of expecting "birds singing and fairy dust" compared to the reality of an enormous amount of hard work ahead.

Expectancy effects rule every part of our lives. We compare our success to those of our colleagues, our date to those of our friends, and our marriage to those of others. Like Abe Lincoln comparing his lack of success to Stephen Douglass's, we base our very judgment of our success in life compared to the lettuce or banana we see others get. And that comparison often brings the same elation or depression that we feel. Reality has nothing to do with it; it all depends on what we have come to expect from our past experiences.

Sometimes, just understanding what to expect makes it easier to survive if the worst happens. Understanding the facts about life can sometimes prevent the self-hatred that can result from internalizing the values of society.

Parents see their own kids, who have far more advantages and iPods than our generation could have imagined, and cannot understand why they are unhappy. It has nothing to do with reality and everything to do with what the youth's own society, their fellow students, have led them to expect; the values and emotions their fellow students embed in their minds.

There is far more to the story than just elation-depression effects. If we had simply been honest with Brooke and so many of our youth, then most roller-coaster emotions would probably never have happened. But we are not honest.

We have used a textbook to teach lifespan to our college students for the last eight years. We review a dozen lifespan textbooks each year, trying to find a better one. Each year, most of us go back to the same one we have used all along. I make no secret of why I prefer this book. It is written by a wonderful lady, Kathleen Berger, who has a remarkable honesty in what she tells students—an honesty that might have saved Brooke Shields, Michele Obama, Megan Markle, and millions of women from the pain of guilt and depression in their lives.

It doesn't take much. All she did was tell students the truth. Bear with me until you figure out what this means. Of all-natural conceptions, when the joining of sperm and egg that produce a living, growing cell, some 69% of those conceptions will fail to develop or implant in the uterus properly. Most will be sloughed off in menstruation, and the woman will never know that she was pregnant. Of those who do implant and begin to grow, some 20% will miscarry. Of all conceptions, only 31% will survive to become a living, newborn baby. Nature's abortions are incredibly common. But that is not what people want to hear.

If the doctors who did the in-vitro fertilization treatment on Brooke Shields had been honest and upfront, they would have told her what to expect. If our society were honest and upfront, every child would grow up being taught this. She should not have been talking about getting her hopes up and them being dashed over and over when in-vitro fertilization failed. In-vitro fertilization succeeds only 32% of the time. That is normal with in-vitro fertilization. That is normal in "normal" fertilization. In normal fertilization, only 31% of fertilized eggs survive to become living newborn babies.

Most of us are not like Octomom. Her In-vitro fertilization produced eight babies. But a doctor would have to have implanted dozens of fertilized eggs in order to produce that many babies.

If she had learned an honest view of reality in our schools and known to expect that 20% of second-trimester babies will miscarry, she might not have been feeling things like, "*Why am I being punished? What did I do wrong...?*" Nothing that happened to her was in any way cause for shame or self-blame. Yet, our schools do not teach this; instead, we let it happen by our silence. The recent emotional honesty of many, from Brooke Shields to First Lady Michelle Obama, to Megan Markle shows how great the failure of our educational system has been.

In college, we go over many causes of birth defects, from genetic ones to anoxia and physical trauma. We talk about avoiding smoking, drinking, excessive vitamin A, and drugs while you are pregnant. We discuss many things we hope to avoid like lead (as in the recent furor over the lead paint on Chinese toys) and mercury (as in some tuna fish), and many more things. Yet, we have no control whatsoever over the overwhelming number of things that cause birth defects and miscarriages. I do not want any student to go away from my class thinking that if they have a miscarriage or a baby born with a birth defect that it is their fault. That would be an awful burden to carry through life—but we do dump that voodoo curse on many young women by our failure to tell the truth in our schools and the media.

Of all babies born, about 3% will have serious genetic defects. By the end of the first year, another 3% will be found. Another 16% will have some genetic defects of varying degrees of seriousness (such as juvenile-onset diabetes or any of the 6,000 known genetic defects). All of us have some genetic defect—a birthmark, spot on the skin, or the shape of our nose. None of us are immune to genetic defect risks. And it does not matter one binary digit whether you are American, Armenian, Russian, Chinese, Iranian, or a Cocker Spaniel—the same percentage of all babies born will have serious genetic defects.

I see too many students who have had a miscarriage or a child born with a birth defect who come to believe that this is some form of a "mark of Cain" or punishment by God for some real or imaginary sin.

Others may come to believe it happened through a fault of their own for having smoked a joint, drinking too much at Spring Break, or even had "bad thoughts." Nonsense. The real feeling of guilt belongs to a society that fails to tell the truth to young people in our public schools.

We are a society that allows our youth to grow up believing nonsense because we are afraid to tell them the truth about how few conceptions survive. People are offended by teaching students something that they do not believe: parents would be upset, Rush Limbaugh would shriek in outrage, committees would be formed, the press would hear of this, the sky would fall.

"How dare you teach this to our children?!" could become the complaint of the year. School principals would cower in fear.

Our society strongly believes in holding people responsible for their behavior. We fire negligent pilots. We make examples of students who behave recklessly and irresponsibly. But there is no one to hold society responsible for far greater negligence: our failure to tell young people the truth.

Simply telling the truth about the causes of birth defects and miscarriage is enough to relieve the feelings of depression and guilt that have caused so many problems in people's minds. And you can read the relief on their faces; "I'm not to blame. It isn't something I did. It's even 'normal.'"

Education can be more effective in relieving depression than psychotherapy or antidepressants. And that, if you think about it, is profound!

Brooke Shields was a beautiful woman with a more than a successful career. Yet, in her own words, she was unable to see anything except to narrowly focus on her own problems. The very fact that the amygdala, the center of fear and anger, the Fight or Flight response, sits right next to the end of the hippocampus, the part of the brain that consolidates short-term into long-term memory, ensures that anything we worry about will remain in our memory, popping up again and again. In the past, this was often useful; today, it is often what hijacks our mind and creates our pain.

The limbic system forces our conscious mind to dwell on our problems. In the old days, that might have been an advantage. It meant that when we were hurting from hunger or afraid of attack, we would concentrate all our attention on planting more food or building our defenses.

Today, the very parts of the brain that ensure we pay attention to problems has turned against us. The things that cause us problems today often cannot be solved, like the fear of what other people will think of us. Yet, the brain forces us to concentrate our attention on even the most trivial problems. The brain itself has become our worst enemy. It is not a mental illness; it is a normal way our brain reacts to today's stresses.

Everyone has problems. Who doesn't? George Bush was President of the United States, the most powerful man in the free world. Did he not have problems? He may not have known it, but he did. Bill Clinton was also President of the United States for eight years. Did he not have problems? Most of those eight years saw a weekly, sometimes daily humiliation in the press over him and Jennifer Flowers, Paula Jones, and Monica Lewinsky, as well as the blue dress, and the stain on the blue dress, and the cigar... Donald Trump is, well...

Look at it this way: no matter what you do in your life, you will probably never have to go through the kind of daily public humiliation that Bill Clinton went through... unless your name is Tiger Woods, Jessie James, Charlie Sheen, Rob Lowe, or any drunken Hollywood celebrity that the media decides is more important to cover than serious news. If they can survive all this and recover, you can also survive and do well.

And the media puts out all these bits of dirt on celebrities because they know that negative emotions will make your brain hijack your conscious mind and force you to pay attention.

Put your problems into perspective. Think about this. What was it you were worrying about this time last year? Or five years ago? Are they the same things you worry about today? Probably not. And a year or five years from now, you will still have something to worry about, but it will probably not be the same thing you are worrying about now. There will always be something to worry about.

We all have problems in life, but that does not keep us from worrying about them now. Most of those problems will solve themselves, one way or another. All our worry and pain will dissipate. New problems will arise. You may rest assured that getting through today's problems will not end tomorrow. The trick is to learn how to cope with those problems and always be aware that most will not last.

THE POWER OF RETRAINING: Counterconditioning

Football superstar Michael Vicks served time in prison for running a dogfighting operation. Pit Bulls were tormented until they were angry and forced to fight, and made more aggressive. Some of them were used as "target dogs" to allow the fighting dogs to develop their self-confidence by attacking dogs who could not fight back. All the dogs were grossly abnormal and unable to function in normal society. Some were too aggressive, constantly fighting with other people or dogs. The dogs who had been the target dogs, who had been beaten repeatedly by the aggressive ones, were cowed, afraid to look up, move, or eat when a human or dog was around. They were psychologically depressed.

No one had any hope that these dogs could be rehabilitated. A spokesperson for the Humane Society called them among "the most aggressively trained fighting dogs in the country," and, along with PETA, the People for the Ethical Treatment of Animals, recommended the dogs be put to sleep.

Instead of doing this, these dogs were taken in by organizations that specialized in rehabilitation, such as a program called "Bad Rap" that specialized in saving Pit Bulls. Using basic psychological principles, they were conditioned to behave appropriately in the real world. They succeeded so well that many of them were successfully adopted; out of 47 selected for the program, 25 were put in foster care, and 22 went to a sanctuary with experienced personnel. One became a certified therapy dog working with kids—this required intensive counter conditioning and desensitization.

If we could do this with dogs, we can do this with people. Sadly, we do not do this when we dump soldiers back into the community. We do not do this with ordinary people who come in for psychotherapy, either. We do not do this for children who have been bullied. Instead of intensive care, we give them a pill and a little talk therapy. If a child is bullied, we tell their mother to "get them help" as if it were their fault, as if therapy could prevent more bullying, and do little else.

Something is wrong with our culture. Why don't we make it better?

Our society creates psychological problems. And they continue because of our failure as professionals, teachers, parents, and society to teach our young people and ourselves an honest understanding of how their minds work, to anticipate the problems we will face before they do, to understand how our behavior affects other people's minds. Most psychological problems are created by the very ignorance and superstition that our society forces upon adults and children's minds, by the value society places on gold and paper, on love, beauty, and success.

We believe in those stories as a society, and we support and pay homage to them, passing them on to our children with the same fervor and mindless devotion as those in Elspeth Huxley's *Flame Trees of Thika,* believing in the magical stories that cursed the mind of the black magic victim. The impact of these beliefs on the mind is total, and the only cure is education.

Only by learning how these forces impact our lives can we go beyond the forces in the environment to control our own lives. It is never as easy as it sounds, but it can be done.

THE CRISIS IN AMERICA THAT KILLS 47,000 AMERICANS EACH YEAR

Every year in America, an average of 5,000 teenagers and young adults commit suicide. Each year, some 47,000 Americans of all ages die from suicide. Some 570,000 suicide attempts are reported every year, and many go unreported. This speaks of an enormous amount of pain and suffering throughout this country, suffering that is never acknowledged in the press and never makes the headlines.

How many Americans have even heard of these statistics? Almost none. How can 47,000 Americans die every year, and yet that fact is not front-page headline news, day after day, month after month?

The only time you ever hear the news media talk about this serious issue is once a year in a 15-second sound bite so that they can claim they covered it, or when something bizarre like a 14-year-old commits suicide over cyberbullying.

America's news media slavishly follows the advice of Joseph Stalin, who is credited with the mantra of our free press. Stalin said, *"…if one person dies it's a tragedy. If thousands die, it's just a statistic."*

To be fair to the press, they cannot cover stories about 47,000 suicides each year because they must deal with far more serious issues, like whether Britney Spears is wearing panties, whether Michael Jackson's brain was buried with his body, or what new politician, who campaigned on family values, got caught fox hunting in Argentina, or who Trump trashed today.

How many people died in the World Trade Center attack, where terrorists hijacked airplanes and drove them into the Center, bringing down the twin towers? Just under 3,000 Americans. And because of that, we invaded Afghanistan and Iraq, spending over a trillion dollars, where some 4,300 American soldiers and over 92,000 Iraqi civilians died, based on the United Nations tally of what happened in Iraq. Terrorism and war dominated the news daily. We did all of that because of one attack that killed 3,000 Americans. Yet 5,000 teenagers and young adults commit suicide every year, and no one even hears about it? The media has declared that it is not news, and that is the end of the discussion.

We could easily save half of those 5,000 teens and young adults if we cared enough to try, just by using the educational system to educate all children about the need to stand up to bullies for their fellow students, how peer group pressure affects us all, and what we need to know to survive in the real world.

We cut back on services for mental health care during the Iraq war. The Mental Health Mental Retardation agency, where I worked, had just hired five new counselors to deal with mental health problems. All had to be let go because the State MHMR budget was ten *million* dollars in the hole. Soon, the total state budget went to 14 *billion* dollars in debt, and they were firing teachers and public workers as well as cutting back on funding for agencies that help others. Even today, when we are doing better, no one wants to spend tax money to deal with serious problems.

Politicians have learned that they can promise to pay people money if they vote for them by giving them a tax cut, so taxes are never raised so that we can help our kids.

I attended a seminar in Texas, where staff from across the state came to discuss the seriousness of this issue. Everyone from psychologists, psychiatrists, social workers to nurses, police, and yes, the Texas Department of Corrections were represented. The Texas prison system is the number one provider of mental health services. We rank 45th out of 50 states in per capita spending on mental health care. The issue came up of how to get the message across to the press as to how desperate the situation had become.

One psychologist suggested we all write our congressional representative, but that has been done to death—politicians only care about sensational issues on which they can be elected, like crime, drugs, terrorism, and reducing your taxes. If they cannot promise to pass laws getting tough on crime, drugs, aliens, or terrorists, then there is no one to focus on the wrath of the public by promising to get tough on others if we vote for them. If they cannot accuse their political rivals of being "soft" on crime, drugs, or terrorists, then they have no emotional issue to use to trash their opponents. If they cannot promise to pay voters for voting for them by giving them a tax break, they are not interested. So, nothing got done.

Another psychologist suggested we picket the news media until they condescended to cover this issue. However, psychologists are way too lazy and afraid of looking "unprofessional" to carry signs; and what would other people think if we acted like teenage war protesters? So, nothing got done.

The best suggestion was the idea of thirty psychologists chained together naked inside the Capitol Rotunda in Austin. That is an idea stolen from PETA. The People for the Ethical Treatment of Animals got no attention at all from the press until three women from PETA showed up wearing nothing but fur coats and took them off. The press said, "Wow, now there is a real news story!" And that got the message out about what their issues are. But I am too old and fat, and nobody wanted to see me naked. So, nothing got done." The same group has the same meeting every year. Every year, they discuss the same issues. And every year, nothing gets done.

The group that met to discuss mental health called itself "Mission Possible," a takeoff on the movie "Mission Impossible." They might just as well have called it "Mission Improbable".

How amazing that it takes something as extreme as PETA did to get the press's attention.

Even the press has a joke they tell about themselves: "If a dog bites a man, that's *not* news. If a man bites a dog, *that's* news!"

CHAPTER 11

LIFE CRISIS

OUR HOPES AND DREAMS

At some point in each of our lives comes a crisis of mind, a questioning of the deepest values of our life, of the hopes and dreams that drive our lives. We may run bang against a crisis of confidence or be troubled by self-doubt or loss of hope for a future. These crises are sometimes called the Mid-Life crisis, though it may happen at any age. Consider Albert Einstein's thoughts, the world's most renowned physicist, on the feelings of his youth. Einstein writes:

> *Even when I was a fairly precocious young man the nothingness of the hopes and strivings that chase most men restlessly throughout life came to my mind with considerable vitality. Furthermore, each of us is condemned by the existence of his stomach to participate in that chase which, in those days was more carefully covered up with glittering words and hypocrisy than is the case today.*

Such a chase, Einstein noted, might satisfy the needs of our stomach; we have to eat to live, but it could never satisfy the mind's needs. At some point, in some way, most of us come face to face with the futility Einstein's chase describes. When we are forced to face that futility late in life, it brings to head the terrible fears that somehow, we have wasted the best of our life, like Lincoln at the age of forty. That perhaps we have spent our youth chasing after goals that have no real value; that what we thought was of great value has turned into *"glittering words and hypocrisy."*

ADULTS WHO TURN ON, TUNE IN, AND DROP OUT:
If you ask teenagers in school if they like school, the majority will say they do. But, if you pin them down, it is not the school they like; it is a chance to gossip with their friends. Some even are happy to get out of a home life that is contentious at best and love to go to school, a safe haven. Students know you need a degree to get a job, but you rarely find any who wants to go to school to get knowledge.

Adults are no different. If you ask adults if they like their job, most will say they do. But, if you pin them down, it is not their job they like; it is a chance to gossip with their friends at work and make some money in the process. Many are happy to get away from a home life that has gone stale or contentious.

In his classic *Walden*, Henry David Thoreau wrote of the great masses of people, *"living out lives of quiet desperation."* "Simplify, simplify..." became his mantra. He left civilization and went to live on the shores of Walden Pond. He lived off beans and potatoes he grew himself. He had no boss. He had no worries about being judged by others. He avoided the entrapments of social society by opting out of what he saw as a confusing world, where he felt he had no role. He also avoided the number one cause of psychological problems—other people.

After years of living on the Pond, he went back to civilization and wrote about his experiences. He made enough money off of *Walden*, that he no longer had to work for a living. In America in the 1960s, "flower children" tried to emulate his experiences with the Hippie movement by living together in idealistic communes. Most eventually went back to civilization, too. It is hard to replace social interaction, TV, and video games, with birds, flowers, and trees, not to mention the hard work of laboring in the dirt to make your own food. Yet, even today, many of us yearn to escape the value judgments of others, the pain of life. Or to replace the pettiness of human behavior with a simpler ideal. One of Thoreau's most famous quotes is still important:

"IF A MAN DOES NOT KEEP PACE WITH HIS COMPANIONS PERHAPS IT IS BECAUSE HE HEARS THE MUSIC OF A DIFFERENT DRUMMER. LET HIM STEP TO THE MUSIC HE HEARS HOWEVER MEASURED OR FAR AWAY."

HENRY DAVID THOREAU, *WALDEN*

Many years ago, an editor in Massachusetts took me to see Walden Pond during the American Bicentennial. I had always wanted to see what Thoreau saw. It was nothing like I expected; it is no pond, but a huge lake. I could hardly imagine what Thoreau saw in his day as I stood in between the Henry David Thoreau Burger and Hot Dog Stand, next to the Henry David Thoreau Paddle Boat Concession. How things have changed.

Our first confrontation with the quandary of Einstein and Thoreau may come when we first look back on the fire of our youth. We remember all the hopes and dreams of our early years. In the flower of our youth, each of us swears a determination to live our lives to the fullest, not to get caught in the system like our parents.

In the later years of life, we look back on the haunted longing of our youth for fame and fortune, and love and glory. We compare the reality of what we have become to the hopes we had in our youth. Somewhere along the path, something got lost. Contrast and comparison, the very basis of how the mind and science work, become a problem when we cannot distinguish between the ideas that are real and those that others have embedded in our minds.

The reality of our life is a violent contrast with what we thought we would be. Now life is half-gone, and reality has set our mold hard. We now think we have gone as far as we can ever go. We will never have the job we wanted, and our marriage will not recover the thrill of youth. Never again can we have that hope of the glory yet to be.

When we were young at school, each hour rang in a new beginning, each new grade marking a rite of passage to a higher level. Now, the fifty-minute hour is gone. There are no more bells to toll the change. The years are not marked by new levels of achievement. We can barely remember what happened last year. The years go by so fast. We may feel our life falling through time. And all that is without smoking pot.

But some of us achieve all we ever hoped. The accolades of life by which we mark our territory are full. Certificates from great institutions validate our merit. Our business is doing well. We have loved and "won." Our children are married. Yet still, we think to ourselves, as in the Peggy Lee song: *"Is that all there is, my friend?"* There must be more to life than *this*! *This* is what I've given half a life to achieve? We come face to face with the "glittering words" that have satisfied only our stomachs.

Something is terribly wrong.

We can see what is terribly wrong in one simple statistic. Every year, over 570,000 Americans are treated for suicide attempts. The national association for suicide says the figure should be closer to 1,400,000, as many are not reported. Vastly more Americans feel hopelessness, guilt, or worry, yet would never even attempt suicide. Over half of all Americans will "seriously consider" suicide at some point in their lives, even though few will ever actually attempt it. This speaks profoundly of a heavy burden of pain and suffering that is largely covered up by the happy news and celebrity trivia we see on our daily version of television reality.

The goals of our youth are not what we expected life would be. They may satisfy our stomachs, but not our minds. They do not ease worries and fears or the feeling that we have wasted our lives. They give us no hope for the future.

LONELINESS IS AN EPIDEMIC
Caused, in Part, By Our Social Values

AARP, the American Association of Retired Persons, wanted to see how much more lonely old age was compared to youth. The results, released in 2011, surprised almost everyone. The loneliest were the individuals between ages 43 and 49; some 43% were lonely. However, only 25% of those over 70 were lonely. How can that be?

It makes no sense that people in mid-life, at the most productive time, are lonelier than the elderly; they are even lonelier than the young. Most of these people are married and working, surrounded by others, yet still lonely.

Being alone in a crowd is far worse than being alone by yourself. We have known this for a long time. But why? When we are surrounded by others and yet still feel alone, we tend to compare ourselves to other people who seem happy and enthusiastic, much as Lincoln and Rick Springfield have. That comparison makes us feel worse. It is a psychological "expectancy effect," a perceptual contrast between what we think we should be and how we see our self as being. When we are alone by ourselves, there is no such comparison, although we may miss the support. But many who are married and working still get no support, and the comparison with others can be haunting.

More than this, our culture dumps a value judgment on those who are alone. We feel sorry for people who are alone, or we judge them as less desirable, maybe even as having a personality defect. All these value judgments are negative emotions associated with the idea of loneliness in our minds. We come to judge ourselves by this idea that society implants in our minds.

This makes sense only when you consider what Charles Horton Cooley called the "imagined judgment of the other mind." If we are at a party by ourselves, we feel alone; in part, because we feel judged by others and found wanting. If we are left out of the office banter, we feel judged by others by being ignored. Even if others never make this comparison, we judge ourselves in our own minds. This tells us something very profound about loneliness; it is a creation of our learned judgment of the other mind, not from any real biological imperative.

Yet there is no doubt that having another to share life with helps make our life easier. Having someone to share good times with, makes the good times better. Having someone to share the bad times with, makes it easier to make it through the bad times.

The negative judgment that others pass on being alone is embedded in our minds. This judgment is what makes us feel worse when we are alone in a crowd. Once we retire, we leave behind much of the comparison that makes us feel lonely. And we no longer have to worry about what our boss thinks.

THE BANNER OF OUR EGO: SUCCESS AND LOVE

"Success and failure are largely self-defined in terms of personal standards. The higher the self-standards, the more likely will given attainments be viewed as failures, regardless of what others might think." Albert Bandura

Two major themes in American society dominate our minds and motives, says sociologist Geoffrey Gorer in his analysis of *The American People;* two patterns that run through the fabric of American lives. Those goals are:

> To be a success.
>
> To be loved.

Yet these are far more than just goals; they are the standard on which we hoist the banner of our egos. They are the criteria by which we judge our lives to be "good." For our life to be "good," we believe we must be a *success* (at dating, sports, interpersonal relations, jobs, raising our children, etc.) And we must be *loved* "for ourselves."

We judge the happiness of our lives against the benchmarks of success and love.

However, there is another side to the coin that no one ever seems to mention. Anytime a society trumpets something as the glory of its ego, it creates an equal and opposite fear.

When we glorify success, we create a fear of *not* being a success, a fear of failure.

When we glorify love, we create a fear of *not* being loved; the fear of being alone.

The result is that the two greatest problems in life for the majority of Americans revolve around losing a job and a loss of love; failure, and divorce are things that make us feel worse about ourselves. These fears are created by society itself, not by any biological imperative in our DNA.

SUCCESS AND FAILURE:
The Carrot and the Stick

More importantly, we will see what has helped others go beyond the problems of youth and what makes for success and failure in our children and ourselves.

The knowledge we need to survive and do well in life is learned. As adults, we can learn these ideas ourselves. As parents, we can teach this to our kids.

The goal of success society dangles before our mind's eye creates an equal fear of failure. Early in the last century, every American was told that this is the land of equal opportunity; that anyone could make it on his own if only he were not lazy and good for nothing. The unstated implication was that if you did not make it on your own, there must be something wrong with you.

CBS News commentator Eric Sevareid in the PBS series *America* chronicled the impact this had on the individual when the great depression first hit in 1929. Sevareid describes how, as a youth, he watched his next-door neighbors, who had been fired from their jobs, get up and go to work every morning dressed for their job as if nothing had happened. Later in the day, he would see the same people dressed in their business suits on street corners downtown selling apples or begging for dimes.

These people were deeply ashamed that they were out of work; they were afraid of what their neighbors and relatives would think of them. It did not occur to people that the fault was with the system or that the depression itself was to blame. They had been brought up to believe that anyone could make it; that everyone except slackers and incompetents could find a job. They did not feel it as a failure of "the system," but as a deeply personal failure of themselves.

The story is that people who lost their fortune or were wiped out in margin calls began jumping from tall buildings. But in 1929, the suicide rate never changed. The stock market started to go up the next year. People said, "Now is the time to get in, don't get left behind." Then, the market crashed again. As it started going up again, new hope arose. And then... it crashed again. Not until 1933 did it reach its lowest point. By that time, people had lost hope. They started jumping out of windows; 1933 had the highest suicide rate in American history.

This is what happens with psychological depression. A single incident will not spark a suicide; we still have a glimmer of hope. Today, those who commit suicide do not do so for no reason. They do not do so because they are weak; they do so because they have finally lost hope.

Most important to understand is that those who commit suicide experience an average of four times more negative events over the last year than the average person. And they blame themselves.

The depth of the personal tragedy is evident in the following quote from a study carried out in the great depression of the 1930s by Zawadski and Lazarsfeld. These are the personal feelings of a 43-year-old mason who had worked all his life and was suddenly unable to find a job, it is an echo of the feelings of Mark Twain, mentioned earlier, after he was fired from his job:

> *How hard and humiliating it is to bear the name of an unemployed man.*
> *When I go out, I cast down my eyes because I feel myself wholly inferior. When I*
> *go along the street, it seems to me that I can't be compared with an average citizen;*
> *that everybody is pointing at me with his finger. I instinctively avoid meeting any-*
> *one. Former acquaintances and friends of better times are no longer so cordial. They*
> *greet me indifferently when we meet. They no longer offer me a cigarette and their*
> *eyes seem to say, "You are not worth it, you don't work."*

He had been labeled in his own mind. Stuck with the terrible onus of being an "unemployed" man, he was "different." It was as if he had been labeled with the curse of "cancer" by his doctor or "mentally ill" by a psychiatrist or psychologist.

It is the same feeling expressed by Mark Twain when he was fired by a newspaper for speaking out against police brutality. He avoided meeting others on the street, *"...I felt meaner and lowlier and more despicable than the worms..."*

It may be difficult to imagine today that people blamed themselves and saw themselves as failures for losing their jobs in the middle of a depression. It was our society's version of the Haitian voodoo curse. Even President Hoover refused to give government money for soup kitchens because he believed that giving people a handout was un-American and morally wrong. He believed that people should have to "stand on their own two feet."

It was "Root hog, or die."

More recently, Lee Iacocca, the extraordinarily successful chair who "saved" Chrysler, shared his feelings after being fired from Ford Motor Company after getting into an argument with a relative of the original Henry Ford. After thirty-two years as an employee, *". . . now, suddenly, I was out of a job. It was gut-wrenching."* He was already wealthy enough that he never had to work again, but he did not stay unemployed for long. Months later, he was made president of Chrysler and brought it back from the brink of bankruptcy. Six years later still, he wrote in his autobiography, "Even today, their pain [of his wife and kids] stays with me . . . I'll never forgive (them)." Lee Iacocca suffered the same feelings, but he did not blame himself.

None of this is limited to the unemployed. Even those who have worked successfully at their jobs for half a lifetime may suddenly be confronted with what some call the mid-life crisis. The Great Recession that began in 2008 left nine million Americans out of work two years later, and an additional nine million having to take a part-time or lessor job. It is a confrontation with a feeling of the meaninglessness and trivia of our lives that has suddenly become apparent.

Even among those of us who have a good job, we may judge our success and find it wanting. Our mind is hit full force with what Einstein called "the nothingness of the hopes and striving" of our lives. It is the sudden feeling that we have spent our lives in pursuit of useless goals. That we have led a life dictated by the fear of what other people would think, and what they had already decided for us. Later, we will see how others have dealt with the same problem.

One point should not be allowed to pass; those who are caught up in the maelstrom of the crisis often fail to realize just how arbitrary the goals by which we judge our lives are, just as those who had lost their jobs in the great depression came to judge themselves unfairly. The value judgments our society imposes on us become the judgments we pass on others, and those judgments we passed on others we come to pass on ourselves.

Their measure of success or failure was wrong. It was founded on the imagined judgment of "what other people will think." Unfortunately, that is not always wrong; other people *do* think that way.

When we are in mid-life our values are dominated by our job. When we get older it is the value of our family that becomes our focus.

TO BE LOVED:
The Harlequin Hunger

Long before the earliest flowering of puberty, the entirety of our society's attention is directed toward the glorification of love. Television pushes it, preachers praise it, and endless balladeers set its essence to song. Yet everywhere, people are suffering the aftereffect of its myth.

Love is the title of a best seller by Leo Buscaglia. Our society trumpets its message, "Love conquers all." "All we really need is love," say the Beatles, although it did not turn out so well for Sir Paul; that part was not in the song. Love makes the world go round, we are told. Books and ballads tell of affairs of the heart. Love stories are a cottage industry.

Love becomes our young's prime goal. Not because they "need" to be loved, but because they have been told the stories of how great it is. They see the thrill of their friends when they find a boyfriend or girlfriend, no matter how big a jerk they turn out to be, and none of them want to be left out in the mad grab for the great secret of life. They have seen their peers' enthusiasm in the adolescent passion of the rating-dating game. Everybody says love is wonderful. It must be so.

Yet, there is much the young have not been told. Surveys have found that as many as 40% of high school students do not date. The 60% who do date experience varying degrees of success... and failure. Yet, if you are one of the 40% who does not date, or even of the 60% who do with what is judged to be less than success, our society always makes you feel as if you are the only one who loses in life's roulette. This fear of being left out, of missing the great chance in life, is perhaps the single most overwhelming fear of adolescence, and it continues into adulthood.

Countless ballads by recording artists put these fears to music in tributes to the heartache and heartbreak of lost love. From folk songs to rock and roll and country-western music, the loss of love has long been a predominant theme.

One of the most successful and poignant of those songs was penned by Janis Ian; it went to number one on the hit parade. I can only assume that it went to number one because her emotions are all too common to people in our society. Her song stands as a memorable tribute to the fear of not being loved that cuts so deep in our youth:

> *I learned the truth at seventeen*
> *That love was meant for beauty queens; For brown-*
> eyed girls with clear skin smiles,
> *Who married young and then retired.*
> *But those of us with ravaged faces,*

Lacking in the social graces,
Desperately remained at home,
Inventing lovers on the phone,
Who called to say, "come dance with me"?
And murmured vague obscenities,
At ugly girls like me, At seventeen.
There must be something terribly wrong . . .
With me.

There is "something terribly wrong..." but not with adolescents. The problem is created by society. We have glorified the emotional feeling of "love" without considering the consequences. The other kids in school have elevated it to a status of divine expectation without ever considering the harm it does. We have not given them the skills, knowledge, or understanding of what is going on in their minds. They cannot cope with a problem they did not create. They cannot escape the consequences of its Romeo and Juliet anguish that we failed to prepare them to understand.

The problems of youth, and the pain and anguish of love, are not caused by any lack of love; nothing of the kind. It is caused by the fact that, like the unemployed fellow in the great depression, our society has made them feel as if they are somehow unworthy, as outcasts and failures, as if there were one great hope in life, and they have somehow missed it. More than this, they have been made to feel that it is somehow their fault, that they have failed without even knowing why, that there "must be something terribly wrong . . . with me."

Because they have never heard anything about other people's lives, they come to feel that they must be the only ones in the world who feel this way. Out of countless high school myths glorifying love, they have heard little that would teach them any other way of seeing reality. We have given them only "glittering words and hypocrisy," not reality.

Adults are sacrificed on the altar of love too. Reality is not always living happily ever after; between 40 and 50 percent of all marriages end in divorce. Nor is this an artifact of the "new morality" exclusive to the young. In Sun City, Arizona, a city noted as a retirement community, the divorce rate has been at an incredible 60%.

This was the generation believed that marriage is forever. What happened? While one or both worked, the couple were apart maybe 50 hours a week. They may have toughed it out for years, but after they retired, and they are together 24/7, the little problems in interpersonal relations become major ones. The irritating grains of sand were magnified by its daily grind. It is no plague of youth, although they get the blame.

Despite one popular myth, we cannot blame divorce rates on newfangled morality coming from universities, either, since college-educated couples have a slightly lower divorce rate than those with less education—not enough to get cocky about, though.

We still cherish the myth that those who stayed married are all living happily ever after. Yet many stay in unhappy marriages because they fear what others would think if they left. Others do so because their religion forbids them to divorce. Still, others stay married "for the sake of the children." Some fear having to go back to living in the rat race of the rating-dating game. Many fear the economic hardship of divorce. They can see no way out. For many, it is often just "comfortable."

"Being in love" carries with it tremendously powerful feelings. The feeling of belonging at last, of knowing there is someone who cares, someone to listen to your hopes, to share your pain and pleasures, and the joy of living, of having found someone who feels you are someone special. There is no denying the powerful positive emotions of love.

Yet, there is something terribly wrong with our society's concept of love; the greater the addiction to culture's illusions, the greater the distress of withdrawal when it does not work out. The harder society pushes the illusions, the more we shatter when we fall.

A friend of mine, a successful businessperson, twice married, twice divorced, gave it all up at the age of forty and went out looking for the "good life." At least he went looking for something he had never quite found. In an honest moment over beer and pretzels, he opened up and described his feelings:

"All my life, I've felt something was missing because I have never been 'in love.' I had heard everyone talking about how great love is, about all the emotions, but I didn't feel it, I didn't have it. I figured it must be me."

What he finally realized, the thing that finally "cured" him, or at least ended his worry, was that there was nothing wrong with him; other people were not better off. If anything, he was more successful and well-adjusted than most. He had been caught up in the unrealistic expectations created by society's glorification of love.

So, what causes such problems? In part, it is a failure of our society's definition of "love." We artificially define love as all things bright and beautiful. All the bad feelings associated with love—jealously, heartache, the problems of getting along—are all defined as something other than love. But what about the "wounded male ego," "hell hath no fury like a woman scorned," and "we only hurt the ones we love?" What about the simple problems of living with another person? These are never mentioned in the high-school version of love; even adults rarely talk about them.

Can we pretend that "love" is all things good, and anything else is not love? Not honestly. We have given a name to all good feelings. We call it "love." But saying that love is all wonderful feelings does not make it true. We have been skewered by our definition. We have failed to give people an honest and accurate understanding of life; we have allowed society to embed unrealistic expectations in ours and our children's minds.

Today, marriage is changing. What we would like to believe about marriage being sacred no longer matters in the reality of living. The fastest-growing family type in America is now the single-parent family. In a single generation, the number of married couples who decide not to have children has doubled, from 10 to 20 percent. Oprah Winfrey was asked in her television interview if she was ever going to marry Steadman, the man she was living with. She replied, *"no"* in no uncertain terms; that marriage is fine if that is what you want, but it was not for her. Just twenty years ago, she would never be allowed to say that on television without others criticizing her.

Marriage often sets up expectations of certain roles and behavior that can damage the relationship. Oprah was honestly aware of that. The cultural ideal of marriage was considered too sacred to allow any other possible interpretation of reality.

If marriage works for you, great, but it will not work for everyone.

Problems are also caused by that gnawing illusion, partly created by the tendency to hide our problems; that other people are better off than we are, and more, that somehow there is something wrong with us if we don't have what we imagine they have.

Yet each of us also knows that other people are sometimes better off than we are... that we could have made more of our lives if only we had the social skills, knowledge, and chances. "If only I knew then what I know now! I could have been a better parent. Our relationship could have worked out. I could have succeeded better at my job..." This is the refrain of every stage of life. Yet we do little to pass on what we know to our children or each other.

Glorifying love is no answer. We have failed to learn the lessons of reality, and we have failed to give our children the simple living and social skills, the understanding of others, and the knowledge it takes to succeed at living.

Knowledge helps.

TO BE OR NOT TO BE: SUCCESS AND LOVE

Each of us, whether prince or pauper, wants to be a success and to be loved. We want to "live life to its fullest." And none of us want to be left out in this bare pedestrian world, in the mad grab for success.

Out of the countless modern-day Horatio Alger myths, the stories about the poor boy who, through hard work and diligence, became a success, and the endless stories of heroes and heroines who live happily ever after, each of us strive to become a success and to be loved.

Our society sets up these goals to be glorified in the same way that medieval society had glorified the quest for gold and glory in conquistador stories. Not just Pissarro and Cortés, but countless stories of conquest and daring filled youth's mouths and minds of Europe. Yet Cortés, like Columbus before him, died a poor and broken man. That was not mentioned in the stories that spoke of glory.

Shakespeare wrote of the effect of such stories, of how young men felt compelled to defend their honor, to make their reputations, to be heroes in battle:

> *Full of strange oaths, and bearded like the pard,*
> *Jealous in honour, sudden and quick in quarrel,*
> *Seeking the bubble reputation Even in the*
> *cannons mouth.*

And like the bubble of glory that marked the conquistadors' quest, the bubble of success and love has left an enormous amount of pain and discomfort in its wake.

Society has led us to feel that we must make a success of ourselves. And what is the image of success projected on our minds? We are fed a constant stream of heroes—famous people, Hollywood stars, politicians, football and war heroes. These are people made famous by the conditioned emotions of applause and the accolades of others. They have their names in the newspapers. Are these the role models of success? Is this what we need to fulfill our lives?

The Civil War general William Tecumseh Sherman went through a long series of bloody battles that saw the death of his colleagues and their glorification as heroes by the press. Sherman later commented, "*We know what fame is. Fame is to die in battle and have your name spelled wrong in the newspaper.*"

Simone de Beauvoir, at the age of sixty, wrote with the sadness of the *"vast miscomprehension,"* the distance between what people think of as a success and the reality of what the successful feel. After winning five gold medals, Michael Phelps lost much of his enthusiasm and says he spent a year doing nothing. He finally got therapy and says it helped greatly. Eric Berne spoke of the balance sheet, a time of agonized reappraisal of all we have achieved. Brooke Shields spoke of *"...why me? Why am I being punished?"* at the high point of her success and fame. T. S. Elliot commented on the "Hoo-ha's" they feel in bed at night with their success over and no goals left to goad them on.

One of the strongest, most self-confident men on earth would have to be a newsman from CBS's "60 Minutes" television show. Mike Wallace, tough, honest, experienced. He exudes strength and confidence.

Decades of success and dealing with hard reality have made him tough. He is sort of the John Wayne of the news profession, not the kind of man who would ever feel depressed or worthless. Yet, the late Mike Wallace went through three major bouts of depression in his life and had the honesty to talk about his feelings.

One of the worst major bouts of depression came when he covered a tough story about General Creighton Abrams, the man who led America's forces in the Vietnam war. It was a scathing indictment of Abrams' conduct during the war, based on the testimony of others who had been there. And it turned out to be wrong. Abrams sued CBS and Mike Wallace. Wallace describes how his hands were shaking so badly in the courtroom that he could barely hold a glass of water.

Although Abrams settled for an apology from CBS and Wallace, so it cost them little, this event had a profound effect on Wallace. His self-confidence was shattered. He said after that, he questioned everything he did. He had built his life and reputation on being a thorough and fair journalist. Now he felt that everything he did was in doubt. He went into a severe depression. He took to reading about depression and said this education was very useful to him. He sought medical treatment, which also helped.

If this can happen to a man as self-confident, successful, and tough as Mike Wallace, it can happen to anyone at any time. Depression does not happen for no reason. It does not happen because of bad genes, except for a small percentage of those who do have a genetic predisposition. It does not happen because of some mythical biochemical imbalance. It happens because of specific events in our lives.

Air Force Colonel Buzz Aldrin, a fearless combat pilot, became one of the most celebrated and renowned astronauts in world history when he touched down on the moon. He had a doctorate from MIT, and a lifetime of success most men could never hope to achieve. Yet shortly after his greatest accomplishment, after being a national celebrity for the first lunar landing, he went into a serious depression, crying, with a feeling of hopelessness. He now had no goals, no future. At the age of forty-one, he began again to find new goals.

Consider the fame of Marilyn Monroe, Janis Joplin, Michael Jackson, and others in contrast to their eventual ends. Fame does not bring peace of mind. Something is terribly wrong. Caught up in a world of unrealistic expectations and the expectations of others, they had never learned to control their own mind.

We glorify soldiers in combat as heroes; reality is quite different. CBS news broke a story that, in a single year in 2005, some 6,250 American veterans committed suicide. It was loudly criticized by the VA as untrue, but they finally had to admit it was accurate. Today, more of our soldiers are dying by suicide than in Afghanistan, Iraq, or Syria combined. Twenty veterans a day commit suicide. And it happens year after year. Even more dramatic is the fact that some 30% of all homeless people in America are veterans.

How can this be happening? How is it possible that no one ever hears about it? Why is this not headline news, day after day? Despite the fine reporting by CBS on veteran suicides, the story did not get a tiny fraction of air time that the "Balloon Boy" fraud received or the months of coverage of who President Trump trashed that day. Other networks did not pick up the stories of suicide; it disappeared within days. How can that be?

We cannot solve the problems of society if we continue to trumpet trivia in our news while we ignore the most serious problems of our culture.

All this was going on as politicians in Washington were using patriotism as a club to beat down their political opponents by accusing them of "not supporting the troops" if they did not support funding the war. What were these critics doing to support the troops after they got home? Not much. As one student, a Marine from the Gulf War, put it, *"When we were in the Marines, they were your family. Everyone was expected to sacrifice for your buddies, for the Corps. When you got out, they just dump you back in society. You never hear from them again."*

In the military, we start in boot camp to break down the individual's self-confidence. We call them "maggots," we yell at them and treat them like dirt as individuals; then we build up their confidence in the Corps; they become a member, not an individual. We train them in the military to be blindly obedient, to kill the enemy, then we dump them into society with no consideration for the fact that we have made so many of them unable to fit in, and we are surprised that they have one of the highest divorce and suicide rates of any profession.

CHAPTER 12

STUDIES OF GENIUS

In the early 1920s Lewis Terman of Stanford University began one of the most extensive studies of genius done yet. For all its impressive title of *Genetic Studies of Genius,* there was very little that was "genetic" in his findings. Even the title came from the assumption, then common, that genius must somehow be inborn.

In the most substantive study of genius ever done, Terman followed some 1,528 Californian children whose IQ scores were in the top 2% of the nation. They are still being followed to this day. His findings started an inquiry into the nature of genius and intelligence that has never ended.

Terman found that 31% of these children had fathers who were professional men. An additional 50% had fathers who were semiprofessional or businessmen. Yet, only 11.8% had fathers who were in skilled labor, and only 6.8% had fathers in the two lowest categories of semiskilled or unskilled professions.

This is a dramatic finding because it means that 81% of the fathers of these "genius" levels came from the top *two* occupational levels (out of *five* levels), compared with only 14% of the total population at that time.

Why? Was it because the genes of doctors, lawyers, and businessmen are more likely to produce children of "genius"? Or was it because these children of the well-to-do had the "better" environment: the advantage of going to better schools and parents more interested in the values of education? Or were there still other factors at work?

Terman's studies never were able to answer the question of genetic influences. But there were strong hints of powerful forces at work in the environment. Terman found that of those children at the highest level, they had access to an average of some 450 volumes of books. Now, even if you averaged child's comic books and science fiction thrillers, you would find it takes an awesome amount to equal 450 volumes.

Having a library of 450 volumes does not mean that the child is likely to read most of them. However, it does say something extraordinary about what that family considers to be important. Few families value education to that extent. Few children are exposed to such an environment.

Terman found that the gifted started early. Nearly half had learned to read before school age, and what they read was dramatically different in quality and quantity from what most children their age were reading.

Terman states, *"... the typical gifted child of seven years reads more books than the unselected child reads at any age up to fifteen years."*

Nothing, not one single finding in the entire fifty years of the Terman study, stands out with such overwhelming importance as that fact.

Even today, *reading* is the single most important ability that differs "genius" from the average. No single ability has ever been found to make as much difference. If you want your child to score high on an IQ test, encourage him or her to read early in life.

Think about that—the gifted child is reading as many books at age seven as the average child at age fifteen. There is no biological gene for the love of reading, no way. The majority of people have read for only the last one hundred years of our history on this planet. And this is not something children can be forced to do easily. It is something that can only occur because those children have found a deep and abiding enthusiasm for reading or knowledge.

Those same parents who fill their houses with books seem to have taken the time to give their children an enthusiasm for the unique knowledge found only in books... the way some parents give their children an interest in sports.

"Early ripe, early rot," was the phrase once applied to children who show early promise. It is common to find people who believe that those children who began with a head start somehow burned out before they became a success. After following these gifted children for over a quarter of a century, Terman found that the early start they were given did not wither. Nearly 70% graduated from college. This group produced ten times as many college graduates and 1,000 times as many Ph.D.'s as the average.

Contrary to the popular myth of genius as being limited, with serious problems, just the opposite was found. Those in the top 2% were healthier, more successful in virtually every way, and even had fewer dental caries. However, much of this was not due to their high IQ, but to the same superior environment that produced a high IQ.

DOES AN IQ TEST REALLY MEASURE INTELLIGENCE?
Or, How to Raise Your IQ 15 Points Without Even Studying!

Anyone can raise their IQ 15 points, to go from Normal IQ (85-115 pts) to Superior IQ (116-130 pts.) or from Superior to Very Super IQ (131-145), what used to be Genius level (145 pts.). All you have to do is take the WAIS IQ test that was used in the 1970s, and your IQ will go up 15 points automatically.

If you want to raise your IQ score by 20 points, just take the Stanford-Binet IQ test used by Terman in the 1920s, and your score will jump 20 points.

How can this happen? Because our score on an IQ test has gone up simply due to our generation having learned more information—from TV, Sesame Street, PBS, toys that teach reading and arithmetic, more awareness by parents of the need to help their children learn, and more test-taking savvy.

The biology of our brain has not changed one iota; the only thing that has are our experiences in our environment. Despite the claims of identical twin studies that IQ is inherited, there is simply no more compelling evidence that it is not based on biology. But neither David Wechsler (WAIS, WISC tests) nor even Alfred Binet, who first developed the test, ever claimed that they measured biological intelligence; that was an illusion that others have created, by the media that wants to claim everything is in our DNA and biologists and psychologists who did not know any better. Both

IQ test inventors cautioned against assuming that these were absolute; they noted that many factors, such as education, experience, and environment, could influence the test score.

This is not to criticize the value of the IQ test; it is still the bread and butter of psychology. I have given perhaps 1,000 such tests in my career and consider it to be a very good way of understanding what a person's ability and problems may be, but it is not a measure of biological intelligence.

Even more of a problem with the assumption that it measures intelligence is the fact that we know for certain what puts individuals in the bottom two or three percent, and that is brain damage. The damage may be genetic (Down's syndrome, Klinefelter syndrome, or any of 6,000 known, identifiable genetic disorders) or the 15% above that (75-84), where people may be affected due to anything from lead or mercury poisoning to educational deficiency.

In other words, we know for certain what biological problems put people in the bottom range of IQ, but there is absolutely no evidence that would show a *biological* cause that puts people in the top two or fifteen percent.

What do we know about the top 2%?

AMERICAN MEN OF SCIENCE:
What Makes for Success in Science?

A further clue to elements in the environment that go into the making of their success comes from a study of the *American Men of Science* by Anne Roe. She found that an unexpectedly large percentage of leading scientists were the sons of professional men, as Terman had been. She further found that these scientists grew up in a home where...

"... for one reason or another learning was valued for its own sake. The social and economic advantages associated with it (learning) were not scorned but they were not the important factors. The interest of many of these men took an intellectual form at quite an early age. This would not be possible if they were not in contact of some sort with such interests and if these did not have value for them. This can be true even in homes where it is not taken for granted that the sons will go on to college."

Quite simply, they grew up in homes where knowledge was valued for the sake of knowledge.

By themselves, these sketches of scientists' early lives are strong hints of powerful forces at work in the environment of the child of success. It is not always easy to separate the environment from genetics, but we must deal with the known forces that we can identify and not the possible forces we cannot. To do this means we must look at studies suggesting what has and has not worked. One area that allows us to separate genetic from environmental forces are studies of birth order effects on success.

THE PRIMACY EFFECT:
The Experience of Being First

The interest in birth order was picked up by the press at the dawn of the space race when an article in *Newsweek* magazine reported to the public that 21 of the first 23 astronauts to travel in space were either first-born or only children. Later, studies of *Who's Who in America* and of similar sources of successful people also showed a greater number of first-born than later born. The difference is nowhere as extreme as the astronaut sample would suggest, more like a 15% advantage, but it is quite striking. To this day, psychologists vigorously debate whether the findings are real or simply an artifact of other variables, and there are many that influence success.

Not that all is sweetness and light for the first-born. Psychologist Stanley Schachter found that first-born children also tend to be more anxious than their younger brothers and sisters.

All this brings up the question of why first-born children would have any advantage at all over later born. There is no way that genetic variables could account for such a difference. There is absolutely no biological advantage to being first-born. The laws of chance that determine the role of the genetic dice at the moment of conception are identical, regardless of your birth order.

Only one possible explanation is left; if first-born children are different from later-born children, it must be because they are treated differently than later-born children.

How could this be? How could two children grow up in exactly the same family, with exactly the same parents, and exactly the same socio-economic environment possibly be different? After all, they would all have exactly the same experiences.

The logic of this argument seems compelling. If they grow up in the same family, they must be exposed to the same experiences. But the logic is misleading; it does not happen that way.

The similarity of experiences is an illusion. No two children ever have exactly the same experiences, even growing up in the same family. And the experiences they do have in common may come at different ages or different moments of susceptibility in their lives. Consider the following:

Parents have a unique reaction to the first-born child. It is known, with very little attempt at humor, as the "new toy effect." It walks, talks, and wets its pants, and most of all, like a good computer, it is infinitely programmable.

Consider the first landing of men on the moon. Most people who saw it on television still remember it; we remember the name of the first astronaut, but what about the third landing, or the fifth or the seventh? Our attention wanes. We are less intent in our interest.

The first-born is the first at everything; the first to have colic or the mumps, the first to take its toddling steps to the encouragement of daddy, the first to go off to school, to the fears of mom, the first to date, to drive, to leave home. And the reaction we give to those who are first, differs greatly from that we give to the second, third, or fourth.

Ooooh, look at baby walking," says mom as she holds baby up by its arms.

"Atta boy, come to papa," says daddy as he cheers him on.

By the time the second-born comes along, it's, "Say, have you noticed junior is walking already?"

And by the time the third-born comes along its "Hey, watch that kid, he's eating dirt again."

The same exaggerated reaction may greet the first-born at many things. When they are first to bring home grades. The first to be in a school play. The first to try out for the school team. The first to be out past curfew. And each new event may be greeted with nervous anticipation or glowing encouragement.

The first-born is the proving ground for the parent's ego. If the child is a success, it is their diploma, their proof that they are worthy of the title of "good parent."

Not that all is for the good. First-born children are more likely to be kept under tight rein, to be controlled and directed as well as helped. Parents, because it is their first such experience, are more likely to overreact to even simple things, like colic or fever, by rushing the child to the doctor. They are more likely to worry over how minor things may affect their first-born. By the time the second-born comes along, they should be feeling more secure and less anxious over the trivia that seemed so important the first time they experienced it. And the anxiety they feel may, in turn, influence the child's feelings, although the evidence for that is not clear.

Surveys show that rules for dating and driving a car are likely to be more relaxed for later-born children. Parents may also be more likely to allow early dating or driving for the second-born—unless, that is, they had a bad experience the first time around, then they may tighten up. But the experience with the first-born should make them less anxious the second time around.

The success of the first-born carries with it the glory or stigma of success for the parents. They are more likely to be anxious that the firstborn succeed. And they tend to go far out of their way to ensure that this happens.

The first hard evidence of this comes from a study by Hilton of how differently mothers react to first and later born. Hilton took sixty children and their mothers and gave the children a series of tests to complete. He watched how the mothers reacted as they watched their children perform.

Twenty of the children were "only" children; twenty were "first-born," and twenty were children with older brothers and sisters.

The children were given puzzles to work out on their own. Their mothers were told to remain seated nearby. When Hilton called an intermission in the testing, the mothers were left alone with their children.

Instead of remaining seated, eighteen of the mothers got up and coached or helped their child practice the test. Fifteen of those eighteen mothers were mothers of first-born or only children.

Out of the sixty mothers, twenty-seven mothers (out of a possible 40) of firstborn and only children helped them directly or encouraged them to practice during the intermission.

Yet only five mothers of later-born children did so.

The implications are strong. Mothers of first-born children are more likely to be anxious and concerned that their child does well.

Mothers who already have a first-born seem to be less concerned about their later-born doing well.

Hilton also found a difference in the way such mothers praised the child's success. Mothers of first-born children tended to express encouragement over the child's success but reduce their encouragement when the child is not correct.

Yet, mothers of later-born children gave about as much encouragement when the child was unsuccessful as they did when the child was successful. It was as if they felt a need to encourage the child but were less concerned with whether or not the child was successful.

THE HARVARD PRESCHOOL STUDY

A massive thirteen-year study by Harvard psychologist Burton White dramatically confirmed Hilton's original finding in the final report of the Harvard Preschool Project, the last of a three-volume series.

White found that mothers spend more than *twice* as much time with their firstborn children as those with later-born children. They spent an average of 40% of the observed time per hour interacting with their first-born compared to only 16.4% with their later-born.

White concludes with this comment on how important such an increased contact may be:

> *... As far as the first few years of life are concerned... experience plays a very substantial role in the rate and level of achievement of early linguistic and cognitive skills."*

Yet, perhaps the most striking finding of the Preschool Project was quite unexpected. White found that simply because the parents *knew* that their child was being tested in the project, the parents dramatically increased their attention toward their children. According to White, virtually every first-born he studied tested as considerably more capable on their third birthdays than such children who have not been included in such a study, *even if they did not receive any preschool training*. That suggests that motivating mothers to spend more time preparing their kids may be as important as any of the preschool activities the children received.

Simply *knowing* that their children were being evaluated resulted in these parents spending a great deal more time helping their children prepare than they otherwise would have. And that fact increased their ability.

At times, I have heard parents in child psychology classes say that they no longer believe "all that stuff" about how the environment affects intelligence. They say that they have already raised one child, and they tried it on the first one, but it "had no effect;" the child just seemed to blossom on its own. Now they doubt that their helping out had any effect.

But the effects of the help they give their first-born are so subtle and take so long to show up that what they do cannot be readily apparent. I suspect that it is this change in their attitude, this feeling that the child "just seems to do alright regardless of what I do," that may result in much of their change in attitude toward later-born children. If parents do not see obvious results, they may conclude that their efforts have no effect. One should not be too quick to dismiss the influence parents have on children, even when there is no dramatic evidence of success. On the other hand, we should not encourage parents to overdo their work. Sometimes, it is a good idea to "let kids be kids."

All of this makes it even more important for parents to be aware of how these subtle differences in what they do or what they avoid doing will affect their child.

The evidence strongly supports the idea that taking extra time to help children learn gives them a substantial advantage, but it tells us nothing about what kind of help we should give. Nor does it tell us what other elements go into making for success in life. That remains to be seen.

THE NOBEL LAUREATES:
The Background of Success

There is a far more astounding finding on the origin of success than anything yet mentioned. Some small social groups of people are consistently vastly more successful than the rest of us. In her review of *America's Scientific Elite*, Harriet Zuckerman found that Jewish people produce a staggering 27% of America's Nobel Laureates.

This fact is all the more remarkable because Jewish people make up only 2% of the American population. Not only that, but there are dramatically more Jewish doctors (23.8%) than would ever be expected from such a small percentage of the population. And there are equally high percentages of lawyers, businessmen, actors, and writers than would ever be expected.

How could such a thing have come to be? How could 2% of our population produce more than 20% of our successful professionals?

The success of Jewish people is even more remarkable, considering the suppression and prejudice that has been common in our past. One hundred years ago, Jews living in Christian Europe were not allowed to teach or even attend public schools. They were not allowed to hold public office, and they were not allowed to be doctors or guild members. More still, they could not even own property. Even Albert Einstein, ten years after his seminal paper on The Special Theory of Relativity, could not get a job teaching at the university in Berlin until his colleagues petitioned the Kaiser for a special dispensation to allow him to teach.

Nor was this was limited to old Europe. In America, even after World War II, American Jews in Hollywood often found it necessary to change their names to make them more Anglo to be accepted or get a job. Some of the most famous actors and comedians have done so, including George Burns (Nathan Birnbaum), Jack Benny, Danny Thomas, and scores more.

Early in the history of psychological testing, new immigrants who came to Ellis Island in the United States were given intelligence tests to see which immigrants were the most desirable. Incredibly, Jews and those coming from non-Anglo Europe found themselves labeled "feebleminded" due to their IQ test scores. It does not seem to have occurred to those giving the tests at that time that people who spoke little English could hardly be expected to do well on an IQ test; they could not even understand the instructions. It is all the more remarkable since today, Jewish people consistently score higher on IQ tests than the national average.

There are only two possible reasons why Jews could consistently perform so much better than the average—superior genes or superior environment. Are Jews genetically superior to non-Jews? It is a reasonable question to ask. However, like most such questions of genetic superiority, the simple truth is that we have no substantial evidence supporting the idea that any group is genetically superior to another.

From what we do know about the nature of the human brain, any idea that one group of people is genetically superior to another seems extremely remote. People seem most likely to believe in the possibility of genetic superiority if we talk about their own race. Simply because it is possible does not make it true; there is hardly a shred of substantial scientific evidence to support that possibility.

That leaves only the environment. What is there about the environment of the Jewish child that makes that child more likely to be successful than the non-Jewish child?

There is an old-world tradition where a drop of honey will be placed on the Torah at Jewish bar mitzvahs. The boy will kiss the honey, which symbolizes the "sweetness of knowledge." This is more than a quaint story; it is basic to understanding the environment of the Jewish child. In a world where they had been made outsiders, where even the guilds and crafts were closed to them, they found education to be the one tool that made them successful. The already existing values of knowledge were made more valuable because it worked for them. They quickly realized that education was the key to getting ahead.

Anthropologist Dorothy Lee writes of the Jewish sheto of Eastern Europe. The sheto are small, tightly knit groups of Jews that had been common in Europe until World War II. She vividly describes the special value often attached to knowledge and learning in the Jewish community.

Education was valued for the sake of education, knowledge for the sake of knowledge. It closely fits with Annie Rowe's description of the home life of American scientists to a striking degree. And it is not just Jewish; the same kind of environment tends to be common in scientists and "genius" backgrounds from every group. In case you have not noticed, today, we see increasing numbers of medical doctors and specialists from India and Pakistan. Few Americans are motivated enough to want to do the hard work needed to get a medical degree.

Yet, this should be no more surprising than to find those famous athletes come from families where the father was much more than an average sports fan.

One has to contrast the extent to which these groups value knowledge and education with the disdain much of the community holds. True, everybody *says* education is a good thing, but the support from society is often little more than lip service. It is common to find that the school system itself holds a deep disdain for knowledge. Conforming to the dress code, "behaving" properly, or being polite to the teacher is often more valued than knowledge.

It was not uncommon, even when I was going to school, to hear other children boasting about how they never studied or about how "I really blew it on that test!" And many more will follow suit, trying to see who did the worst. Some will then feel embarrassed at their own success and feel a need to downplay their good grades to fit in; to not be thought a "nerd."

The deep respect, awe, and admiration for knowledge and the reverence toward those who are learned people within the community are distinguishing hallmarks of Jewish history. But there may be other reasons, too.

One of the nation's leading social psychologists, Philip Zimbardo of Stanford University, described what may be one of the most important differences of all—the findings of a comparison between the child-rearing practices of Japan and those of other societies. He found marked differences in some areas.

Post-war Japan has been a miracle story of recovery, from the point of ruin to their current position of virtual world domination in many industries. Automobiles, computers, and cameras, along with robotics and steel production, have allowed them to surge ahead of much of western society. There is also a tradition of respect for knowledge in Japanese society, but there is a more recent addition of harsh college admission standards. Parents start their children early to ensure that they will pass the qualifying exams spread out along the way.

Success is not seen as something an individual achieves on his own. In Japanese tradition, the family and community are emphasized as being more important than the individual. Children are taught that it is a good thing to set aside one's own pride for the good of the group. Western values of the individual who triumphs over the group, as in "Chariots of Fire," where one person is glorified for winning a footrace, are seen as lacking in humility. They may see this as an undeserved form of self-glorification.

The important difference is that when a Japanese child succeeds—if he does well in school or on the job—then everyone is praised. The grandparents are praised for having maintained tradition in raising the child's parents; the parents are praised for having raised such a worthy child; the school is praised for doing such a good job in educating him, and so on. It is not that the child does not benefit from this praise of others; to the contrary, he feels a sense of pride in having made others feel happy, rather than having succeeded.

In our western society, we mete out praise on quite different standards. We tend to praise a child when he is successful. The praise is conditional—children are not praised unless they succeed. And all our society's modeling stories, the heroic tales of those who are successful, tell of how an individual struggled against enormous odds to finally emerge victorious.

In marked contrast to both these societies, Zimbardo notes that a child is praised for *trying* in a Jewish family. Whether or not he succeeds is not the major goal; what is important is that he or she tries.

Zimbardo feels that praising the child for trying, whether or not he succeeds, is what produces a certain degree of persistence. It gives the child a feeling of confidence, a feeling that it is all right to fail. That enables him to keep on trying, even in the face of great adversity.

There is no way to prove that the Jewish practice of praising a child for trying is the one thing that produces such enormous creativity and success, but it sounds like the best child-rearing advice I have ever heard. Nor is it limited to Jewish society. "Try and try again" was a famous Scottish expression and has at least been encouraged in much of society, even if it has not been praised. However, it seems likely to be of great importance.

Japanese culture has made a very productive society. Yet, there is one problem in their educational system that they acknowledge—while they do extraordinarily well at taking new ideas and improving on them, they have had great difficulty in creating new ideas to begin with. Much of the computer, videocassette recorders, and automotive innovations in Japanese products began as western ideas. Often the western culture that spawned the new idea failed to capitalize on its promise, and the Japanese saw its potential and developed it.

The Japanese educational system is deeply worried that they will reach a stalling point because they have not had a history of creating new ideas themselves. Their schools explore new ways of teaching children to develop this creativity, and they have been more eager to seek new changes than western society.

American news reports of the Japanese success story often emphasize the Japanese lack of creativity, and, in contrast, America's advantage over them. But we should not take too much comfort in this. Remember that only thirty years ago, the term "Made in Japan" was considered a derisive comment on the poor-quality copies of American tools and knick-knacks that the Japanese made then. Barely twenty years later, the term "Made in America" became a comment on how poor our quality in automobiles and other products had become compared to the high-quality standards of the Japanese.

Even as this is being written, the Japanese have already taken steps to remedy the "creativity gap." They are industriously studying every method they find to encourage creativity. We should not be too surprised to find that they will very quickly surpass us in this area.

Japanese education has succeeded. In 1990, a comparison of Japanese and American schoolchildren found that the mathematics ability of the top 5% of American high school students equaled 50% of the same level of Japanese students. This is sometimes taken as proof that we should "get tough," require more homework, and force students to spend long hours in school... But this does not tell the whole story.

In Japan, primary grade schools are quite easy-going. Children are not expected to sit in their seats quietly, as are American children. They are expected to explore, play, and experience things on their own... to have fun. It is more a combination of the permissiveness of Summerhill and the laissez-faire of Montessori.

The easy-going approach of the Japanese to learning dramatically changes in the higher grades of school. By high school, every parent and child know that only the best students will be able to get into universities. They realize early on that their scores on a college entrance exam will determine their future. Teachers push students, parents hire tutors, and children begin to worry over not making it into a university.

Therefore, the pressure becomes intense in high school years. So intense that recent studies show a surprising increase in suicide rates among high school students. Instead of the pleasure of learning, fear of failure takes over. The suicide rate has exceeded that of America and South Korea, another very high-achieving nation. Both Japan and South Korea have started to try to reduce the suicide rate. Korea has regulations that prevent students and their tutors from staying up past midnight studying, as they had been doing.

Yet, those who do succeed in getting into the universities find a new world. Japanese universities are not so much institutions of higher learning as they are finishing schools. Here, students are expected to socialize, party, and make business contacts that will eventually result in becoming a member of one corporation or another. And once they join a company, it is often a lifetime commitment.

It is probably not the Japanese educational system per se that is superior to ours. In every analysis, we keep returning to one major factor in producing a society's success—the esteem in which knowledge and ideas are held. Social science professor Ezra Vogel vividly describes the importance of learning to the Japanese in his book *Japan as Number 1*:

> *If any single factor explains Japanese success, it is the group-directed quest for knowledge. In virtually every important organization and community where people share a common interest, from the national government to individual private firms, from cities to villages, devoted leaders worry about the future of their organizations, and to these leaders, nothing is more important than the information and knowledge that the organizations might one day need... But these leading circles were merely articulating the latest formation of what had already become conventional wisdom, the supreme importance of the pursuit of knowledge.*

The quest for knowledge does not end with graduation from school. Japan publishes twice as many books and newspapers that are purchased per person as in America. Each graduate is expected to continue to be a student; to learn and continue his education. Being a student is respected, and everyone from homemakers to corporation presidents is encouraged to broaden their education. Vogel notes:

> *Study is a social activity which continues throughout life. By the time Japanese youth complete formal schooling, not only have they acquired general information, but they have acquired the habit of studying in groups. Even if they read alone, they discuss their reading with peers. University education may be more important for certification than learning, and the social atmosphere may impede probing, but it does not impede groups of students from continuing to learn...*

Public educational television is well-funded in Japan; that says a great deal about what their society values. In contrast, American public educational television had a golden age in the 1970s, but funding was cut back in the 1980s. The Reagan administration's avowed intent was to end what they saw as "controversial" programming. Whatever the cause, the innovative and controversial programs were replaced by staid, uncontroversial trivia. Once-prime stations such as PBS dropped their imaginative programming in favor of shows on antiques and great British baking. No new series of the caliber of "Cosmos" was introduced for a very long time. Symphony orchestras and operas were considered "highbrow" enough to pass for education. Forty-year-old black and white movies that would never be shown on a network were labeled "classics" and liberally used as fillers to kill air time. Knowledge was ignored. Controversy was censored. There were always a few exceptions, but the exceptions became few indeed. Public television disgraced itself in public.

PBS has only begun to recover. One of the few exceptions was the Ken Burns' series on the Civil War, which touched on some brilliant new views of what the reality of the Civil War was—far better than anything taught in public schools. This is the kind of programming that should be routinely used in our high schools as an adjunct to the boring, tasteless textbooks. Or instead of the textbooks. If we combined this with a DVD, and periodic test questions they could all answer on a computer, this would be a winning combination to get knowledge across to students. We are all visual learners, and far more of this would make an impression than words on a page. It would increase motivation, interest, and memory.

Because the government stopped supporting public educational television, Public Broadcasting became dependent on selling thirty-second advertisements to big corporations and begging for handouts from its viewers. American educational television died in the 1980s through a lack of funding, fear of controversy, and neutering their most creative people. History must record that its death was presided over by the same politicians who loudly claimed they valued education in each election campaign. PBS made something of a comeback in the 2000s when they reprised COSMOS with Neil DeGrasse Tyson, but elsewhere there was nothing even mildly "controversial."

Elsewhere, the Discovery Channel ran shows on Shark Week and survival programs became a fad. The History Channel stopped showing programming on history and settled on back-to-back shows of Ancient Aliens. Other networks went out for Ghost Hunters and Bigfoot and bizarre claims of pseudo-scientific information.

The Japanese success story is remarkable. That does not mean that we need to imitate all their problems and pressures, however. We need to use what is worthwhile and throw the rest the hell away.

And what is worthwhile?

News reports of Korean and Vietnamese children who come to America and surpass American children in the educational system are becoming common. One of the extraordinary similarities between Jewish, Japanese, and other Asian success stories is in what their parents emphasize and why. What do they teach their children to value? What do they have in common?

The answer is perhaps not surprising, but it is too subtle to be obvious. Every one of these groups learned early on that the only key to success in our society was education. Having a college degree is the only chance they had of getting ahead in their new culture. The best jobs are only open to those who have advanced degrees. For those who do not, the job market can be very lean.

That simple fact led their parents to emphasize the value of knowledge, to teach the importance of education to their children. A college degree is their admission ticket for success in life. Education is what makes it possible for a Vietnamese child, who could barely speak English, with no other hope but to grow up to be a grocer, to leap over others in society, and to "make something of themselves." One Vietnamese child grew up to be a psychiatrist I worked with—that would have been almost impossible without the value of education. Most could not hope that their father would let them have a job in his corporation, that a friend would give them a promotion, that they would receive some special favor, or, like Donald Trump, that their dad would give them 412 million dollars. Their only hope was education; the college degree was what opened doors for them.

Education is what made them into doctors, lawyers, physicists, teachers, and Nobel laureates. Have you noticed that so many new doctors are from India, Pakistan, and other countries? American students just want to be entertained; knowledge and information are secondary to sports, socializing, and Facebook.

Increasingly, our society closes jobs to people who do not have college degrees. Businesses want to hire people with degrees for their best positions. The irony is that, even here, the value of the degree is not so much for the education it represents as for the fact that those who manage to make it through college have jumped through the hoops; gone over the many hurdles; they have demonstrated drive, purpose, and an ability to follow through on a goal.

Motivation is in high demand. Corporations want the hurdle jumpers; they feel that those who "toughed it out" and survived through college will be self-starters, work without prompting, not be quitters, and more likely to succeed at their company.

The worst recession since the Great Depression of 1929 began in December of 2008. It is often called the "man-cession" because most jobs lost were male—in the automobile and manufacturing industries, as well as Wall Street and elsewhere. It has been estimated that as much as 50% of those jobs will not return. The jobs that prospered through the recession were ones that required a degree—nursing, medicine, and teaching. These are the fields that added jobs, while other areas lost out.

In the recession that began in 2020, the world of work was different. Women were hurt the most during the pandemic when schools closed and nine million women were out of work and having to care for their children on their own.

All of this will change for the better, but a college degree is likely to return to being the most important thing the average person can get, to improve their status in life.

CHAPTER 13

IDEALS IN AN AGE WITHOUT IDEALS

We all believe in some very high ideals. Thomas Jefferson wrote some of those ideals into the Declaration of Independence: "*We hold these truths to be self-evident, that all men are created equal, endowed by their creator with certain inalienable rights, among those, life, liberty and the pursuit of happiness.*" I always thought it was remarkable that Jefferson included in those rights "the pursuit of happiness."

Jefferson kept slaves on his Monticello plantation when he wrote those words. Many Americans who could afford slaves also kept slaves, including President George Washington, whose slaves supported his whiskey distillery—the largest rye whisky distillery in the Americas, and who grew hemp on his farm—for rope, of course. Washington later freed his slaves.

Apparently, it was not so self-evident that all men are created equal. Apparently, too, the people we label "heroes" were quite different from what we consider "heroes" today. If we like someone, we censor the bad about them. If we dislike someone, we trumpet the bad at high lung. This bias is deeply embedded in our educational system, news media, politics (watch the political attack ads), and ourselves, which makes it hard for us to understand how our own minds really work.

The positive or negative emotions associated with these stimuli change our very perception of reality. The mind takes its cue from the emotions first embedded in our brain by our culture, peers, and educational system. We judge others and ourselves based on those cues, and the judgments come to control our minds.

We teach a much-censored version of reality in our public schools. In part, because of that censorship, we never learn to question the value judgments we pass on other people. That is why it takes so very long to change this country for the better.

Thomas Jefferson was smart enough to be aware of the vast discrepancy between his ideals and the reality of his time. He freed his slaves and provided for them in his will. Unfortunately, he died in great debt, and his slaves were sold to help pay his debt.

The laws of his day demanded that if you were to find an escaped slave, you were required to turn him over to his owner. To fail to obey the law would have been a crime, the same as harboring a fugitive, or an illegal alien, is today.

It took eighty years and a Civil War to change the law in America. More Americans died in the Civil War—over 600,000—than in all wars we have ever fought added together. Even then, it took another one hundred years and the visual images of police losing dogs and water cannons on Martin Luther King and his followers to bring the issue of civil rights to the President and Congress, and longer still, for many people to catch up. Women were not mentioned in the Declaration; they were not even allowed to own property in many states. It took over one hundred and forty years before women got the vote.

How astonishing; that an entire nation would go to war for five years, killing over 600,000 men, yet none would sit down and discuss the issue rationally. To them, it was all about *"stand up for your beliefs. Fight for what you believe in."* Those are the ideas still glorified in our culture. No one ever says, "Why do you believe that?" or "What is the evidence?" This lack of understanding is basic to most wars and disagreements between nations, individuals, spouses, bosses, and politicians.

It is never easy to see the other person's side of the issue. No one wants to admit they are wrong; no one asks, "What is the evidence?" No one asks, "Is this worth the pain it brings?" There is no more powerful lesson in human history than that. Yet we studiously ignore the enormous fallibility of human judgment.

Most of us believe in some very noble religious ideals, "Love thy neighbor," or "Love thine enemy," or "Bless him who curses you." Do we bless those who curse us? No; we curse them right back.

Do we, "Love thine enemy?" Hardly. We rarely even make the effort to try to understand what their point of view may be.

We believe in the ideal, "Judge not, that ye not be judged." Yet we live in a world in which we constantly judge others. We do not act on the ideals we say we believe in.

When we judge ourselves, we judge ourselves based on our highest ideals, not on what we actually do. Yet when we judge other people, we do not judge them on their highest ideals; we judge them on what little we see of what they actually do.

The same is true about judging other nations. When Americans judge America, we judge based on our highest ideals, not on what we actually do. But when Americans judge other countries, we do not judge them based on their highest ideals; we judge them on what we see of what they do.

The relativity of our value judgments is an important lesson to teach about life. If we fail to understand that, then we fail to understand something quite profound about the way our minds work. And we fail to understand the value judgments that others pass on us.

We do not teach these insights to our children in our schools; we only teach them platitudes. They grow up believing we already have a democracy because it says so in the Constitution. Yet the constitution of the former Soviet Union was based in part on the American Constitution. It guaranteed freedom of speech and religion. They had neither.

A constitution is nothing more than ink on paper; words attached to emotions and ideas. It has value only to the extent that people are sufficiently well-educated to understand the importance of those values and able and willing to take action to stand up for other people's rights. People are pretty good at standing up for their own rights or the ideas they believe in, but not at all good at standing up for the rights of people we may not even like. The failure to teach these simple insights to our children is an enormous failure of our culture. Teaching only ideals creates the illusion that they already exist, and we have no need to improve our country or ourselves.

To understand the mind of the other person, we must come to understand that we are all a prisoner of our unique view of reality. Every one of us considers ourselves to be a good, fair, decent, and caring individual, but we often do not behave according to our ideals because we do not learn any understanding of them. We are never forced to face the reality of our self-deception in our value judgments, and neither are others when they judge us.

That simple insight goes a long way in helping us understand why others behave as they do toward us, and vice versa. That is something we all need to understand to make it through this life filled with other people; bosses, spouses, friends, co-workers, politicians, and people of different nations—all of whom are looking at us, as we do at them, through their narrow perception of reality. Those who fail to grasp that simple fact cannot understand themselves or others. If we do not understand this, we are at a disadvantage in dealing with reality.

CHAPTER 14

25 POINTS

For thousands of years, our doctors opened people's veins and bled them to make them well. For decades doctors refused to believe the "germ" theory of disease; that tiny, microscopic organisms could kill a big adult. Only after they reached a "tipping point" were they forced to change.

Psychology, education, and psychiatry today are at that same tipping point. The ideas we learned from our textbooks do not deal with the serious problems of life. Instead of trying to understand the more basic cause of the problems of life, we open a vein with talk therapy and labeling. Instead, we must work with the real causes of problems, in our environment. And use our most effective potential tool, education, to fix the problems.

PREVENTION is always more effective than using therapy after problems arise. Preventing heart disease and stroke by lowering blood pressure, preventing lung cancer by educating young people to avoid smoking, are both far more effective than treating these problems after the problems begin. The same is true for psychological problems.

IF WE EDUCATE PEOPLE to understand that others have been through anxiety and depression in life, and gone on to succeed, then it effectively inoculates people against such loss of self-esteem. That kind of preventive psychology is more effective than waiting to do therapy after society has labeled them with negative emotions. Preventive psychology can be far more effective than psychotherapy. But, if our psychologists and schools do not take advantage of this and use it to educate all our kids, then we continue to fail them.

PREPARE LIKE A FIRST RESPONDER. Train. Practice. Make a plan. Learn from the lessons of others that bad things happen to all of us. Use this understanding to prepare for the problems that will come. Steel yourself. Plan how you will react. Learn to use what you now know to remind yourself, over and over, you can overcome.

On the Snowball from hell: Do not become a self-licking snow cone. Stop ruminating about the mistakes of the past. Learn from the example of others that even the best of people go through hell in life.

- **THE MOST EFFECTIVE TREATMENT** for the problems of youth, or adults, comes from what is often the cause of the problems in the first place, the people themselves. People must be educated to understand the effect that they can have on others, for better or for worse, and enlisted as a means of preventing the problems, starting in the schools.
- **THE SECOND MOST EFFECTIVE TREATMENT** for such problems is learning about the problems others have faced. That removes the terrible feeling that they are all alone, that no one else has ever been there, that there is no hope. It is desensitizing to the problems of life. Knowledge provides hope. Learning social skills improves the human condition.
-
- **OUR EDUCATIONAL SYSTEM** could use similar examples to educate all children to help each other to prevent this destructive effect before it happens. The other students are key to stopping this, to changing their fellow students' lives for the better, and changing our culture. Yet, nothing is more predictable than that our educational system will do nothing. So, it is up to teachers and parents to do the best they can.

When therapy and schools cannot help...

Improve yourself. Do not let the ignorance and arrogance of others determine your thoughts. Learn from the problems of life the rest of us go through in life that you are not alone. Become an expert in recognizing the bogus thinking of others and our self-blame.

- We all need a life-coach, someone to share our success with, that makes success better. And someone to share our pain, that lessens the pain of living. This book is about how to become your own mind-coach. Educate yourself. Never stop learning
- The whole purpose of all of this is to improve yourself, gain control of your own mind. You do not have to be a genius, or even a success, at anything except becoming the best you can.
- But we can all learn from the success of others what can help us in life.
- **Keep on keeping on. Keep on when all seems against you. Keep on against the worst life has to offer. Keep on when hope seems lost. Nothing else is more important to understand than that.**

Most of us give up after failing only a few times. Few understand this. That is what makes this simple bit of knowledge so powerful.

- Never think that other people succeed because they are better than you. It is about being able to persist in the face of the inevitable problems of life. And a little luck. And a lot of work. And about learning what you need to know to succeed. Knowledge provides skills.

-

- **COUNTER-CONDITIONING AND DESENSITIZATION WORK**: You cannot just tell people to "get over it." That never works. But, if you *show* them examples of how even attractive, successful people have been called names, put down, and dumped on, then that can reduce their emotional trauma.

THE MOST IMPORTANT LESSON FOR THE CAUSE AND CURE OF MOST PROBLEMS:

COMES FROM A STATEMENT BY PSYCHOLOGIST ABE MASLOW:

"Let people clearly realize that every time we hurt or humiliate or belittle another human being, we become forces for the creation of psychopathology, even if those be small forces. Let people clearly realize that every time we are kind, decent, psychologically democratic, we become psychotherapeutic forces, even if those be small forces."

- # THE MOST IMPORTANT LESSON FOR DEALING WITH ANXIETY
  - *Comes from the lesson learned by Lara Jefferson about anticipatory anxiety:*

"At least I have learned this: Nothing is ever so bad when it is actually happening to us as when we are dreading, fearing, anticipating it. It is the fear we build in our mind that gives a thing the power to cause us greater pain."

Make a plan. Prepare like a First Responder. Repeatedly remind yourself about what you have worried about before, which turned out not to be as bad as you anticipated.

- **ANXIETY AND DEPRESSION ARE RARELY "MENTAL ILLNESSES", although a small percentage may be biological.**

They are the natural reactions of the brain to the mental pain of life. The degree of the problems is determined by the number of experiences and the intensity of experiences that cause anxiety or depression. The "Snowball from Hell Syndrome" can follow a long series of negative experiences, leading the brain to *anticipate*, that the same will continue. Counter-conditioning and desensitization are keys to changing the reaction in the brain.

We know from studies that those soldiers who come down with PTSD are the ones on the ground who have had the most experiences with blood and death. Those who shoot hellfire missiles from a plane, and never see the effect, rarely come down with PTSD. We know from studies that those who commit suicide, to kill the pain of remembering, are those who have had 3 to 4 times more negative experiences over the past year than average. *That contributes to the Snowball effect.* Just understanding that can be a first step to recovery.

- **PRACTICE THE BEST ADVICE:** like an actor rehearsing their role, rehearsing what we say or do, rehearsing our "lines" will make us better at whatever it is we want to do. See the example of Albert Ellis in CH VIII for a clue as to how to do this.

-
-

THE FAILURE OF PSYCHOLOGY AND PSYCHIATRY

THE GOOD NEWS: IF YOU FAIL, YOU WILL SUCCEED

- The good news is that, of more than half a million suicide attempts each year that *fail*, only 10% of males and 3% of females will go on to commit suicide. If we get them help, if we let them know that we care, if we can give them goals and hope, treatment does work. Yet, we never get them help until *after* they attempt suicide.

-

- We could easily save half of the 5,000 teens and young adults that commit suicide each year if we really cared enough to try. But nothing gets done. The only way to get to those individuals is to start where the problems begin—in the school system itself. But that would be too controversial, so nothing gets done.

• THE BAD NEWS: NOTHING WE HAVE TRIED HAS WORKED

-
- The bad news is that in the 1970s, an average of 25,000 Americans committed suicide each year. Now that figure is over 47,000 a year. Some of the increase may be due to population growth. Yet, we now have about six times more psychologists, psychiatrists, and counselors than we had in the 1970s, and also about six times more antidepressant drugs, and we have not made so much as a dent in preventing suicide.

-

- To deal with this problem, we have to start prevention in the public school system where these problems begin, not wait for them to attempt suicide. Educate kids to be better parents than we were. Teach them to understand how we treat each other matters.

ON SUCCESS IN LIFE: POSITIVE EMOTIONS MOTIVATE US: Just As Negative Emotions Inhibit Us

- **KNOWLEDGE THAT EVERY SUCCESSFUL PERSON** in history has gone through problems in life, can be desensitizing. That can help us go beyond the hurdles we encounter, instead of being discouraged by them.

-

- **THE SOCIAL SKILLS** that make it possible to do well in life are essential for every area of life. Interpersonal relations are basic for professionals, bosses, employees, spouses, dating, and even raising our children.

-

- **EVERYONE PRAISES THEIR CHILDREN FOR SUCCEEDING.**

Philip Zimbardo noted that some of the best parents praise their kids for *trying*. Encouragement to keep trying in the face of adversity is valuable.

- **ANY STIMULUS-ANY EMOTION**: Something as terrifying as a thunderstorm can easily be made into something that creates a lifelong feeling of "exhilaration" and "excitement". Something as painful and damaging as sports, the number one reason for children being admitted to emergency rooms, we have made into something that creates a lifetime of "excitement" and "exhilaration" over sports. We could do the same thing with the value of knowledge. We do not.

OUR SCHOOL SYSTEM MAKES KNOWLEDGE PAINFUL AND SPORTS EXCITING. The glorification of sports by our schools, with weekly pep rallies, cheerleaders cheering, majorettes prancing and bands playing, has made sports exciting. The failure of our schools to simply point out the far greater importance of science and knowledge is a critical failure in our society.

- **WE COULD EASILY MAKE OUR SCHOOLS BETTER:**

All we would have to do is stop glorifying sports and show students how the value of science and knowledge have changed medicine, agriculture, technology for the better. Do not just teach the bleached, bare bones of science, teach WHY it is important. It would be easy.

o **Easy.**
o Will we do it?'
o Not a chance.

REPLACE OUR TEXTBOOKS WITH AUDIO-VIDEO COMPUTER INSTRUCTION AND TESTING.

o Computers and programmed learning can replace textbooks. Only a small "Manhattan Project" would be needed, with the teachers who are best able to get ideas across, we can make a revolution in how education works. This is the kind of programming that should be routinely used in our high schools as an adjunct to the boring, tasteless textbooks.

Or instead of the textbooks. If we combined this with a DVD, and periodic test questions they could all answer on a computer, ask questions, show video examples, explain ideas visually, stick to serious issues, do not submerge the students in such a morass of information, and do not ask questions of obscure information just to give you a "normal distribution" of grades. This would be a winning combination to get knowledge across to students.

We are all visual learners, and far more of this would make an impression than words on a page. And yet, we still teach the way Socrates did 2,400 years ago, by talking. Video and programmed learning would increase motivation, interest, and memory. Hire computer gamers. *Make it fun.* Test on the computer every 10 minutes or so. If they miss a question, they go back to the beginning of the previous section. Much like video games where you have to shoot all the bad guys, and if you mess up, you have to start at the beginning of that chapter until you get them all.

SCHOOLS NEED TO TEACH ALTERNATIVES TO SPANKING:

"...if the only tool you have is a hammer, everything looks like a nail." Psychologist Abe Maslow

When the only childrearing tool parents know is the "teach 'em a lesson" cliché, everything calls for punishment.

- **IF ENOUGH PARENTS LEARN THE SECRETS OF HOW TO MOTIVATE THEIR CHILDREN...** we might be able to make it through another quarter of a century of the glorification of trivia.

-

- **MAKE LEARNING FUN:** Spend just 5 to 15 minutes a day to get kids started on reading a book or, gasp, homework. Enthuse over learning. Sports are fun because we do it with other people. We hear the "...roar of the Sunday crowd." Learning is painful if we are expected to do it alone.

USE HOMEWORK AS AN OPPORTUNITY: We make knowledge painful when we expect children to "do your own homework". If parents spent just 10 minutes a day to get their kids started on homework, to enthuse over reading or math. Depending on their age: You read a sentence in a book, let them read a sentence. Ask them what they think occasionally. You do a math problem, let them do a math problem. Do not criticize or judge their mistakes, they will learn with a light touch. After a while, you can even back off and ask them to show or tell you what they did when they finish.

- **KIDS LOVE TO DO THINGS WITH THEIR PARENTS.** Take advantage of this when they are young. By the time they are teenagers, they may not want anything to do with you. Their peer group will own all of their allegiance; another benefit of our school system.

USE A LIGHT TOUCH. One student who was doing this with her daughter said she had to keep correcting her daughter's pronunciation. Finally, her daughter said, "I don't want to do this anymore, it's no fun." They will learn just by listening to you, you don't need to correct them all the time. A light touch is generally best.

- **ENTHUSE OVER THE VALUE OF LEARNING.** But do not overdo praise or enthusiasm. Occasionally is good. Too often dilutes the value of praise or enthusiasm, just as parents or teachers who are too demanding make learning itself painful for their children.

-

- **THE BEST PARENTS MAY OVERDO IT. Don't force it.** It is generally good to be positive and enthusiastic, but that is not how the real world is. Prepare them to function in the real world as best you can.

ON MOTIVATION: How to make something as frightening as a thunderstorm into something that inspires awe and exhilaration. If psychology has a secret as dramatic as the atomic bomb was in the science of physics, then this is it. I expect that the first society that learns the secrets of motivation and education and teaches it to its future parents will have a dramatic economic, social, and scientific advantage over those societies that do not. But that is not what we do in America. Instead, we glorify sports, entertainment, celebrity, personal opinions and heroes—anything but knowledge.

Even work does not have to be drudgery; it can give the child a good emotional feeling—of being a part of an experience, pride in making a contribution, of doing something worthwhile, and happiness in helping out their parents. Thank them for helping you take out the garbage, let them know you really do appreciate their help. That is an emotion far removed from having to do something "because I said so" or "because it is your job." But to do this means we must let them know we appreciate their help—do it with them; do not make it into something they are forced to do alone. After a while, you can back off and let them take over the task.

Parents who spend only a few minutes a day, as Richard Feynman's father did, getting their children excited about reading a good book, asking questions and doing their homework with them, or enthusing over a flower, can make an enormous difference.

- **AMERICA PUTS ENTERTAINMENT AHEAD OF KNOWLEDGE**
-

- **The prestigious PEW Research** group lists America as 24th in Science and 38th in Math. We are behind every industrialized nation on earth. We are a joke to the other nations who once admired us.

-

- **Only a few decades ago**, America was number one in the world in producing college graduates. Now we are twenty-second. Fully one in four American students never graduate from high school—a rate unheard of in the rest of the industrial world. We are now twentieth in producing high school graduates.

-

- **One in five Americans** is "functionally illiterate," unable to do math well enough to compute the gas mileage on their oil-burning SUVs or read well enough to follow simple instructions to screw a Chinese made oil filter onto their Korean-made car or to work the Made in Singapore remote control on their Japanese-made 3-D Plasma Limbic System Tickler.

-

- **AMERICA COULD TURN THIS AROUND, just as we have turned global warming around.** Oooops. Too late. Ignorance and arrogance have trumped knowledge and understanding. Our failure to educate students about the value of knowledge and science, our compulsion to glorify sports and personal opinions, has made this impossible. Much like our failure to understand the exponential growth of CO_2 until global climate change has reached the point of no return. Maybe the news media could save us by telling us what we need to do to turn this around?

- No.

- The best we can hope is that individual teachers and parents can still get across to the next generation what is important in life so that there is still a little hope for change.

THE NATIONS WHO LEARN THE "SECRETS" OF HOW TO MOTIVATE THEIR CHILDREN WILL OUTCLASS US IN THE DECADES TO COME. WE WILL BE LEFT IN THE DUST.

CHAPTER 15

EPILOGUE

Surviving the problems of life is a life-long struggle. The more we learn, the better off we will be.

With over 47,000 Americans committing suicide each year, every year, the toll of unhappiness is high. It all starts with the value judgments we pass on others and ourselves. We could easily save half of those Americans if we cared enough to try.

The news media is fond of saying, "*We don't make the news; we just report it.*" Yet, they are the ones who decide what news is and what they do not want to cover. They have decided not to tread on any area that might upset people. The result, whether in medicine, psychiatry, or psychology, is that the people rarely hear about the evidence they need to make an informed decision. It is all tightly censored by the media and our school system.

PSYCHOLOGY AS AN EDUCATIONAL SCIENCE, NOT JUST A THERAPY

The problem of suicide is only a fraction of a much larger problem; our failure to give teens and adults the knowledge and information we need to survive and succeed in a difficult world has created vast problems in interpersonal relations, marriage, school, sex, the workplace, raising children, and every area of human existence. This book is one small beginning of an attempt to provide some of that knowledge. This book is about the problems that plague us all—the problems of living. But we need a massive project with the best minds in the field, by using film and literature to provide an understanding, to use the school system to get the information across to the people.

We cannot allow an endless stream of Holden Caulfield's to go through life feeling like failures. We cannot let adults come bang into a reality they never expected because they never learned an honest view of reality. We must give our children and ourselves the skills and knowledge needed to stay sane in an insane world.

HOW BAD IS IT REALLY?

Republicans think mental health is a farce, an unnecessary expense that detracts from their being able to give even greater tax breaks to themselves and their wealthy campaign contributors. Why waste money on people who should be able to pull themselves up by their bootstraps? If Republicans can't promise to pay you money if you vote for them, by giving you a tax cut, what chance would they have to be elected?

Democrats think we should throw more money at the problem. Hire more counselors for our schools, pay more money for more people to get psychotherapy. Money is not going to solve the problem.

Psychologists think more money should be spent to go into the schools, find students with psychological problems, label them, and give more money to psychologists to get them help.

If you really want to help, none of this will work. Instead, we should use the existing school system to educate our children to understand how they can make a difference, to learn they are not alone, to *show* them the stories of all of the successful people who have been through Hell and gone on to do well, to teach the skills of living, the lessons of life, and to teach them to be more effective parents than we were by giving them a few parenting skills. And teach them the *value* of science, not just the boring details.

If we fail at dealing with the real problems, with our schools and our media and our childrearing, we will never change a broken system; we will continue to fall farther behind the rest of the world.

CHAPTER 16

BIBLOGRAPHY

Bibliography

Altus, W. D. 1967. Birth order and its sequelae. Inter- national Journal of Psychiatry 3:23-32.

Ader, R., and Conklin, P. M. 1963. Handling of pregnant rats: Effects on emotionality of their offspring. Science 142:411-12.

Adorn°, T. W., Brunswik, E. F., Levinson, D. J., and Sanford, R. N. 1950. The authoritarian personality: Studies in prejudice. New York: Harper & Row.

Allport, G. W. 1954. The nature of prejudice. Boston: Beacon Press.

Aronson, E. 1972. The social animal. San Francisco: W. H. Freeman.

Asch, S. E. 1956. Studies in independence and conformity: A minority of one against a unanimous majority. Psychological Monographs 9:70.

Ayllon, T., and Houghton, E. 1963. Intensive treatment of psychotic behavior by stimulus satiation and food reinforcement. Behavior Research and Therapy 1:53-61.

Azrin, N. H., Hutchinson, R. R., and Hake, D. F. 1967_ Attack, avoidance, and escape reactions to aversive shock. Journal of

Experimental Analysis of Behavior 10:131-48.

Bandura, A. 1969. Principles of behavior modification- New York: Holt, Rinehart, & Winston.

Bandura, A., Ross, D., and Ross, S. A. 1963.

Imitation of film-mediated aggressive models. Journal of Abnormal and Social Psychology 66:3-11.

Bandura, A., and Walters, R. H. 1959.

Adolescent aggression. New York: Ronald Press.

Beers, C. 1934. A mind that found itself. New Yu& Longmans Green.

Belmont, L., and Marolla, F. A. 1973. Birth order, family size, and intelligence. Science 182:10961101.

Benedict, R. 1934. Patterns of culture. Boston: Houghton-Mifflin.

Bernstein, I. 1969. Lecture on primate behavior. Arlington, Texas.

Birch, H. G. 1945. The relation of previous experience to insightful problem-solving. Journal of Comparative Psychology 38:36796.

Breland, H. M. 1973. Reply to Schooler's article on birth order. Psychological Bulletin 80, 3:210-12.

Bruner, J. S. 1962. The conditions of creativity. In Contemporary approaches to creative thinking, eds. Gruber, Terell, and Wertheimer. New York: Atherton.

Bruner, J. S., and Kenney, H. J. 1966. On multiple ordering. In Studies in cognitive growth, eds. Bruner, Olver, and Greenfield. New York: Wiley & Sons.

Burlingame, Michael 2013, Abraham Lincoln: A Life Amazon.

Carpenter, C. R. 1964. Naturalistic behavior of nonhuman primates. University Park: Pennsylvania State Univ. Press.

Carpenter, C. R. 1965. The howlers of Barro Colorado Island. In Primate behavior: Field studies of monkeys and apes, ed. I. DeVore. New York: Holt, Rinehart, & Winston.

Carroll, L. 1946a. Through the looking glass and what Alice found there. New York: Random House.

Carroll, L. 1946b. Alice's adventures in wonderland. New York: Random House.

Chodoff, P. 1970. The German concentration camp as psychological stress. Archives of General Psychiatry 22, 1:78-87.

Clemens, S. L. 1923. The adventures of Huckleberry Finn. New York: Harper Brothers.

Clemens, S. L. 1958. The adventures of Tom Sawyer. New York: Dodd.

Clemens, S. L. 1962. Letters from the earth. New York: Harper & Row.

Collias, N. E. 1944. Aggressive behavior among vertebrate animals. Physiological Zoology 17:83-123

Collias, N. E. 1951. Problems and principles of animal sociology. In Comparative psychology, ed. C. P. Stone, pp. 388-422. Englewood Cliffs, N.J.: Prentice- Hall.

Cooley, C. H. 1909. Social organization. New York: Scribner. Cooley, C. H. 1927. Life and the student: Roadside notes on human nature, society, and letters. New York: Knopf.

Cooley, C. H. 1964. Human nature and the social order. New York: Schocken Books.

Coopersmith, S. 1968. Studies in self-esteem. Scientific American 218, 17:96-100.

Cooley, C. H. 1909. Social organization. New York: Scribner. Cooley, C. H. 1927. Life and the student: Roadside notes on human nature, society, and letters. New York: Knopf.

Craig, Wendy. See "The In-Crowd and Social Cruelty" by 20/20 films re John Stossel 2011

Davis, K. 1947. Final note on a case of extreme isolation. American Journal of Sociology 52:432-37.

Dollard, J., Doob, L., Miller, N. E., Mowrer, 0. H., and Sears, R. R. 1939. Frustration and aggression. New Haven: Yale Univ. Press.

Descartes, R. 1961. Meditations on the first philosophy, trans. L. 0. Lafleur. New York: Liberal Arts Press.

DeVore, I., and Eimerl, S. 1965. The primates. New York: Time, Inc. Dollard, J., and Miller, N. E. 1950. Personality and psychotherapy: An analysis in terms of learning, thinking, and culture. New York: McGraw-Hill.

Espar, E. A. 1967. Max Meyer in America. Journal of the History of the Behavioral Sciences 3, 2: 107-31.

Frankl, V. E. 1963. Man's search for meaning, trans. Ilse Lusch. Boston: Beacon Press.

Franklin, B. 1950. The autobiography of Benjamin Franklin and selections from his other writings. New York: Modern Library.

Freud, A. 1946. The ego and the mechanisms of defense. New York: International Universities Press.

Freud, S. 1920. A general introduction to psycho-analysis. London: Bonji and Liverighreud, S. 1935. Autobiography. New York: Norton.

Freud, S. 1938. The basic writings of Sigmund Freud.The Modern Library. New York: Random House.

Freud, S. 1949. Collected papers, vol. IV.

London: Hogarth. (Also published by Basic Books, N.Y.,1959)

Gallup, Gordon C. 1971. It's done with mirrors-chimps and self-concept. Psychology Today 4, 10:58.

Galton, F. 1869. Hereditary genius: An inquiry into its Origin MacMillan

Goodall, J. 1965. Chimpanzees at the Gombe Stream Reserve. In Primate behavior, ed. I. DeVore, pp. 423-73. New York: Holt, Rinehart, & Winston. Goodall, J. 1971. In the shadow of man. Boston: Houghton-Mifflin Hall, C. S. 1951. The genetics of behavior. In Handbook of Experimental Psychology, ed. S. S. Stevens. New York: Wiley & Sons.

Hansen, C. 1969. Witchcraft at Salem. New York

Harlow, H. F. 1949. The formation of learning sets. Psychological Review 56:51-65. Harlow, H. F. 1958. The nature of love. American Psychologist 13:673-85.

Harlow, H. F. 1962. The heterosexual affectional system in monkeys. American Psychologist 17:1-9.

Harlow, H. F., and Harlow, M. K. 1949. Learning to think. Scientific American 181:3639.

Harlow, H. F., and Harlow, M. K. 1962. Social deprivation in monkeys. Scientific American 207:137-46.

Harlow, H. F., and Harlow, M. K. 1966.

Learning to love. American Scientist 54:24472. Harlow, H. F., and Suomi, S. J. 1971. From thought to therapy: Lessons from a primate laboratory. American Scientist 59:538-49.

Harlow, H. F., and Zimmerman, R. R. 1959.

Affectional responses in the infant monkey. Science Aug. 21, 1959:421-31. Hebb, D. 0. 1949. The organization of behavior. New York: Wiley & Sons.

Heidbreder, E. 1933. Seven psychologies.

New York: Century Hall, C. S. 1941.

Temperament: A survey of animal studies.

Psychological Bulletin 38:909-43.

Hess, E. H. 1959. Imprinting: An effect of early experience. Science 130:133-41.

Hilton, I. 1967. Differences in the behavior of mother toward first- and later-born children New York: Henry Holt & Co. (Later edition,

Harper Torchbooks, Harper & Row, 1961.)

Inhelder, B., and Piaget, J. 1958. The growth of logical thinking from childhood through adolescence. New York: Basic Books. Shaping and specificity to discriminative stimulus. Journal of Comparative and Physiological Psychology 63:13-49.

Inhelder, B., and Piaget, J. 1959. The early growth of logic in the child. New York: Harper & Row.

Iverson, G. R., Longcor, W. H., Mosteller, F., Gilbert, J. P., and Youtz, C. 1971. Bias and runs in dice- throwing and recording: A few million throws. Psychometrica 36:1-17.

James, W. 1892. Psychology: The briefer course.

Jefferson, L. 1948. These are my sisters. Tulsa, Oklahoma: Vickers Publishing Co. Jones, M. C. 1924. The elimination of children's fear.

Journal of Experimental Psychology 1:328-90.

Kelly, H. H. 1950. The warm-cold variable in the first impression of persons. Journal of Personality 18: 431

Kendler, H. H., and Kendler, T. S. 1962. Vertical and horizontal processes in problem solving Prentice-Hall. Psychological Review, 69:1-16.

Kendler, T. S., and Kendler, H. H. 1962. Inferential behavior in children as a function of age and subgoal constancy. Journal of Experimental Psychology 64:460-66.

Kinsey, A. C., Pomeroy, W. B., and Martin, C. E. 1948. Sexual behavior in the human male. Philadelphia: W. B. Saunders.

Kinsey, A. C., Pomeroy, W. B., Martin, C. E., and Gebhard, P. H. 1953. Sexual behavior in the human female. Philadelphia: W. B. Saunders.

Klukhohn, C. 1949. Mirror for man: The relation of anthropology to modern life. New York: Whittlesey House.

Kuebler- Ross, Kathern 2014 Life Lessons:

Two Experts on Death and Dying Teach Us About the Mysteries of Life and Living

Mead, Margaret (1972). Blackberry Winter:

My Earlier Years. New York: William Morrow.

Mead, Margaret. 1977. Audio recording of a lecture delivered July 11, 1977.

Mind of Man, Film: NET Audio-Visual Center,

Office for Academic Affairs, Indiana

University, Bloomington, Indiana, 47401

Monkeys, apes, and man. 1971. National Geographic Television Special. Morgan, C. L. 1894. An introduction to comparative psychology. London: Scott.

Mowrer, 0. H. 1939. A stimulus-response analysis of anxiety and its role as a reinforcing agent. Psycho-logical Review 46:553-65.

Mowrer, 0. H. 1960a. Learning theory and behavior. New York: Wiley & Sons.

Mowrer, 0. H. 1960b. Learning theory and the symbolic processes. New York: Wiley & Sons.

Mowrer, O. H., and Mowrer, W. M. 1965. Enuresis an etiological and therapeutic

Newton, G., and Levine, S. 1968. Early experience and behavior: the psychobiology of development Springfield, Ill.; Charles C. Thomas.

Neisser, U. 1967. Cognitive Psychology. New York: Appleton-Century-Crofts. Newell, A., and Simon, H. A. 1972. Human

Pavlov, I. P. 1927. Conditioned reflexes: An investigation of the physiological activity of the cerebral cortex, trans. G. Anrep. London: Oxford Univ. Press

Pavlov, I. P. 1928. Lectures on conditioned reflexes, trans. H. Grant. New York: International Publishers.

Penfield, W. 1959. The interpretive cortex. Science 129:1719-25.

Perceval, J. Esq. 1840 A narrative of the treatment of a gentleman during a state of mental derangement_ London: Effingham

Wilson. (Later edition, 1961, ed. G. Bateson. Stanford: Stanford Univ. Press.)

Piaget, J. 1930. The child's conception of physical causality. New York: Harcourt, Brace.

Piaget, J. 1948. The moral judgment of the child. Glencoe, Ill.: The Free Press.

Piaget, J. 1951. The child's conception of the world. New York: Humanities Press.

Piaget, J. 1954. The construction of reality in the child, trans. M. Cook. New York: Basic Books. Plato. 1928. The works of Plato, ed. I.

Edman. (Jowett trans.) New York: Modern Library.

Rogers, C. R. 1951. Client-centered therapy:

Its current practice, implications, and theory.

Boston: Houghton-Mifflin. Rogers, C. R. 1961.

On becoming a person. Boston: Houghton Mifflin.

Rosenthal, R. 1966. Experimenter effects in behavioral research. New York: Appleton-Century-Crofts.

Rosenzweig, S. 1943. An experimental study of "repression" with special reference to need-persistive and ego-defensive reactions to frustrations. Journal of Experimental Psychology 32:64-74.

Roth, M. 1957. Interaction of genetic and environmental factors in the causation of schizophrenia. In Schizophrenia: Somatic aspects, ed. D. Richter. New York: Macmillan. Russell, B. 1950. Nobel Prize acceptance speech. Stockholm.

Sackett, G. P. 1965. Effects of rearing conditions upon the behavior of rhesus monkeys. Child Development 36:855-68.

Sartre, J. P. 1964. The Words, trans. B. Frecthman. New York: Braziller.

Schachter, S. 1959. The psychology of affiliation. Stanford: Stanford Univ. Press.

Schachter, S. 1963. Birth order, eminence, and higher education. American Sociological Review 28:757-68.

Scheinfeld, A. 1965. Your heredity and environment. Philadelphia: J. P. Lippincott.

Schlipp, P. A. 1970. Albert Einsteinphilosopher-scientist. Library of Living Philosophers. La Salle, Ill.: Open Court Publishing Co.

Schooler, C. 1972. Birth order effects: Not here, not now! Psychological Bulletin 78:16175.

Scott, J. P. 1958. Critical periods in the development of social behavior in puppies.

Psychosomatic Medicine 20:42-54. Scott, J. P. 1962. Critical periods in behavioral development. Science 138:949-58.

Shenk, Joshua, Lincoln's Melancholy How

Depression Challenged a President and Fueled His Greatness Houghton Mifflin, October 2005.

Sherif, M., Harvey, O. J., White, B. J., Hood, W. R., and Sherif, C. W. 1961. Intergroup conflict and cooperation: The Robbers Cave experiment. Norman, Okla.: Univ. of Oklahoma Press.

Skinner, B. F. 1948. Walden two. New York: Macmillan. Skinner, B. F. 1951. How to teach animals. Scientific American 185, 6:26-9. Skinner, B. F. 1953. Science and human behavior. New York: Macmillan.

Skinner, B. F. 1971. Beyond freedom and dignity. New York: Knopf. Skinner, B. F. 1976. Particulars of my life. New York: Knopf.

Solomon, R. L., Kamin, L. S., and Wynne, L. C. 1953. Traumatic avoidance learning: The outcomes of several extinction procedures with dogs. Journal of Abnormal and Social Psychology 48:291-302

Solomon, Andrew, Noonday Demons: An Atlas of Depression 2001 Simon & Schuster, N.Y.

Stossel, John 2010 20/20 ABC Family Teens: What Makes Them Tick?

Stossel, John 2010 20/20 ABC The In-Crowd and Social Cruelty.

Vonnegut, Kurt Jr. Breakfast of Champions, 1974 by Delacorte press.

Vonnegut, Kurt. 1999.—. Palm Sunday: An Autobiographical Collage.

Wallace, Mike 1969 Playboy Magazine interview, Dec.

Winfrey, Oprah in an interview with Piers Morgan Jan. 18, 2011 CNN.

CHAPTER 17

ACKNOWLEDGMENTS

Editing and formatting by LUNAPEA

Pictures licensed from Depositphotos

Cover assist from Germania

Cover art by agsandrew and photo by rfphoto

Interior photo by ana_om

My profound appreciation for all who have come before.

I have tried to give credit throughout the book.

My profound apologies for any I have missed.

www.ingramcontent.com/pod-product-compliance
Lightning Source LLC
Chambersburg PA
CBHW080801300326
41914CB00055B/1006